AQUACULTURE

APPLIED SCIENCE SERIES

A TEXT BOOK OF
AQUACULTURE

Dr. M. SRINIVASULU REDDY
M.Sc., Ph.D.
Department of Zoology
Sri Venkateswara University
Post-graduate Centre
Kavali - 524 202
A.P., India

Dr. K.R.S. SAMBASIVA RAO
M.Sc., M.A., Ph.D.
Department of Zoology &
Centre for Biotechnology
Nagarjuna University
Nagarjunanagar - 522 510
A.P., India

Editor
Dr. DIGUMARTI BHASKARA RAO
M.Sc., M.A., M.A., M.Ed., Ph.D.
R.V.R. College of Education
Guntur - 522 002
Andhra Pradesh, India

Discovery Publishing House
New Delhi-110002

First Published – 1999

Reprinted – 2025

ISBN: 978-81-7141-482-6

A Text Book of Aquaculture

Published by:

DISCOVERY PUBLISHING HOUSE PVT. LTD.

4383/4B, Ansari Road, Darya Ganj
New Delhi-110 002 (India)
Phone: +91-11-23279245; 23253475; 43596065
Mobile: +91 9811179893 / +91 9871656464
E-mail: discoverybooksindia@gmail.com
orderdphbooks@gmail.com
namitwasan9@gmail.com
web: www.discoverypublishinggroup.com

Printed at:
Infinity Imaging Systems
Delhi

EDITOR's...

Aquaculture came on to the scene as a promising enterprise of economic and nutritional requirements of mankind. Whether in large ponds, sea cages or tiny backyard tanks, aquaculture holds much promise for meeting increasing food demands. In fact, with most capture fisheries in decline, aquaculture is the best way to maintain and increase supplies of marine and freshwater fish and prawn.

Recognising the role of aquaculture, many courses are offered by different types of institutions and extensive efforts are put in by several business firms and individual farmers. As, both, students and practitioners try to harvest much from aquaculture, it is very much essential to have a comprehensive literature on the subject.

The present volume on aquaculture will be of great use to the readers as it explains various aspects of aquaculture in a pleasant way. The knowledge of aquaculture obtained from this work will help in achieving the goals.

This book – ***AQUACULTURE*** – is a part of my series – *Applied Science* – of books for ready use in daily life.

Dr. D. Bhaskara Rao

PREFACE

Aquaculture provides important economic and nutritional benefits to mankind. Aquaculture is overwhelmingly concentrated in the developing world, which accounts for more than 85% of output by volume and 71% by value. Exports of high value species such as shrimp and prawns earn much needed foreign currency for these countries. Most importantly for food security, the production, processing and sale of fish offer the prospects of improved local nutrition by providing a ready source of high-quality protein as well as giving an opportunity to generate income. Small-scale farmers see aquaculture as a way of making their food supply more secure by spreading their risks: pests or drought may decimate their maize or paddy but there will still be fish to eat or trade for other foods. So, nevertheless, the challenge for aquaculture is formidable. Appropriate planning, environmental considerations, proper system management and disease control will have to play a more important role than at present if crashes in production are to be avoided.

For such a vast multidimensional field, aquaculture, a host of books are available from different authors meeting the aquaculture requirements of either the students or the farmers. In the light of the above facts, an attempt has been made by us to bring out a comprehensive book on aquaculture. We have presented the subject in a simple language with suitable examples and illustrations in order to make the readers understand and apply this knowledge in a practical way. We hope that this book will meet the requirements of aquaculture practitioners and pursuers.

Dr. K.R.S. Sambasiva Rao
Dr. M. Srinivasulu Reddy

CONTENTS

1

TAXONOMY

INTRODUCTION

Aquaculture is broadly defined as the farming of aquatic organisms such as prawns, molluscs, sea weeds, algae in addition to various fishes. Aquaculture improves the nutritional standards of the people and also it generates income and employment oppurtunities. Fish culture is in vogue for the past several centuries. The aquaculture in India has become an integral part of the rural development. In India and particularly in Andhra Pradesh, the fresh, brackish and marine water resources are vast. At present India is also one among the fish and prawn producing countries in the world scenario. Similarly inland fish production also is at reasonably high rate. Aquaculture technologies were rapidly developed in India in the fields of induced breeding for the production of large amounts of fish seed of good quality, breeding techniques of prawns, hatchery technology, technology for nutritive feeds and disease control.

Fish occupies an important place in Indian mythology, history and tradition. One of the incarnations of God in Indian mythology was in the form of "Matsyavathara". Fish remains, indicative of their use as food by the people, were obtained in excavations at Mohenjodaro and Harappa of Indus Valley Civilisation (2500 B.C. - 1500 B.C.). The people of Egypt were probably the first in the world to culture fish (2500 B.C.). Fish farming is widespread in China since 2000 B.C. In India fish culture in reservoirs was mentioned in Arthasastra (about 300 B.C.) by Kautilya. Traditional fish culture existing in the eastern states of India particularly Bengal, Bihar and Orissa has gradually spread to the other regions by the end of the 19th century.

SELECTION OF SPECIES FOR CULTURE

Jhingran and Gopalakrishnan (1974) included about 465 species, belonging to 28 families of plants and 107 families of animals, in a catalogue of cultivable aquatic organisms. The main aim of the fish or prawn culture is to produce maximum quantity of fish and prawn. Though several species of fish and prawns are available, only few species are suitable for culture in fresh and brackish waters. Such of the species are selected which are fast growing, non-predatory and efficient converters of natural food resources of water body, both from the point of view of production and economics. Therefore the selection of culturable species plays an important role in fish or prawn farming. In other words the production of a confined water body mainly depends on the selection of species. The criteria for selection of species can be broadly divided into, viz., Biological and Economic.

Biological Criteria

1. Quick growing varieties of species should be selected. The selected species should effectively utilize the natural food and the food should be converted into the stuff of the body.

2. There should be compatibility among the species selected for culture, intended to be stocked in the same pond. Hence species having different feeding habits and behaviour should be selected to avoid serious competition among the different species.

3. The species selected for culture should be capable of withstanding the environmental fluctuations like oxygen content, pH, temperature, turbidity, etc. They should be able to survive under temporary bad water conditions and also hardships in transport. The species should show high percentage of survival.

4. The species selected for the culture activity should be resistant for various diseases.

5. They should be non-predatory.

6. There should be high food conversion efficiency among the selected varieties of species.

7. The seed of cultivable varieties of species should be available in sufficient quantities.

8. The culture species should accept the supplementary or artificial feed along with the natural foods available in the culture environment.

9. The selected varieties of species should be able to breed by the induced method of breeding.

Economic Criteria

1. The species selected for culture should have good market demand.

2. They should have consumer preference.

TAXONOMY OF CULTIVABLE FISHES

Class	Osteichthyes	
Sub Class	Actinopterygii	
Order	Gonorynchiformes	
Sub-Order	Chanoidei	
Family	Chanidae	*Chanos chanos* (Milk fish) Marine and Estuarine
Order	Cypriniformes	
Sub-Order	Cyprinodei	
Family	Cyprinidae	Freshwater fishes *Cirrhina mrigala* (Mrigala) *Catla catla* (Catla) Ctenopharyngodon *idellus* (Grass carp) *Cyprinus carpio* (Common carp) *Hypopthalmichthy molitrix* (Silver carp) *Labeo cali basu* (Kalabasu) *Labeo rohita* (Rohu) *Tor tor* (Mahaseer)

Order Siluriformes	
Family Siluridae	*Wallago attu* (Freshwater shark)
FamilySaccobranchidae	Freshwater fishes
	Hetropneustes fossilis (Cat fish Singhi)
Family Claridae	Freshwater fishes
	Clarias batrachus (Cat fish, Magur)
Order Channiformes	Freshwater fishes
Family Channidae	*Channa marulia* (Giant headed murrel)
	Channa punctata (Green snake headed murrel)
	Channa striata (Stripped snake headed murrel)
Order Mugiliformes	Brackish water fishes
Family Mugilidae	*Etroplus suratensis* (Pearl spot)
	Mugil cephalus (Grey mullet)
	(Marine fish entering brackish water and rivers)
	Mugil corsula (Freshwater mullet)
	Mugil dussumieri (Grey mullet)
	Mugil tade (Green-black mullet)
Order Perciformes	
Family Centropomidae	*Lates calcifer* (Cock-up)
	(Marine fish entering freshwater)
Family Cichlidae	*Tilapia mossambica* (Tilapia)
	(Freshwater fish occurring in estuaries)
Family Anabantidae	*Anabas testudineus*
	(Climbing perch)
Family Osphronemidae	*Osphronemus goramy* (Gouramy)

1. *Chanos chanos*

The milk fish is the most common species in the brackish water systems in South-east Asia. Body is compressed, scales small and toothless, single dorsal fin inserted in front of the pelvis, caudal fin prominent with subequal lobes. Body-bright silvery along the back, metallic greenish, blue on the top of the head.

Fig. 1.1 Chanos chanos

2. *Cirrhinus mrigala and Cirrhinus reba*

These carps resembles Labeo rohita in general shape but is somewhat more slender. The body is elongated and compressed. The mouth is wide and lips are thin. The body is silvery but dark grey along back. The snout is obtusely rounded and it may have some pores. The upper lip is either fringed or entire and it is not continuous with the lower lip. The lower jaw is sharp with small tubercle at the symphysis. Barbels are of very small size and its number may be two or four. In certain cases it may be absent. In *C. mrigala* the upper margin of the body is concave particularly in the posterier side. The dorsal fin has 15-16 rays and the lateral line scales are 40-45 in number. In *C. reba,* the scales are more exposed and it looks larger. Scales have dark edges. Sometimes, bluish longitudinal bands may be present on the body. The dorsal fin is less than body height. Lateral line scales are 35-38 in number.

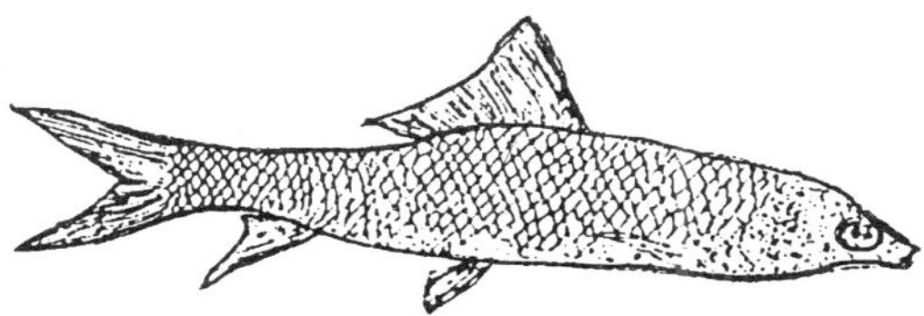

Fig. 1.2 Cirrhinus mrigala

3. *Catla catla*

It is a silvery fish with a pinkish tinge. The body is short with rounded abdomen and the dorsal profile is more convex in comparision to the ventral one. It has wide mouth with prominent lower jaw. Eyes are large and situated in the anterior half of the head. Pores may be present on the snout. Barbels are absent. Upper lip is absent and the lower lip is very thick. Lower jaw has a movable articulation at the symphysis. Dorsal fin is inserted above tip of pectoral fin, and has 17-19 rays. Caudal fin is forked.

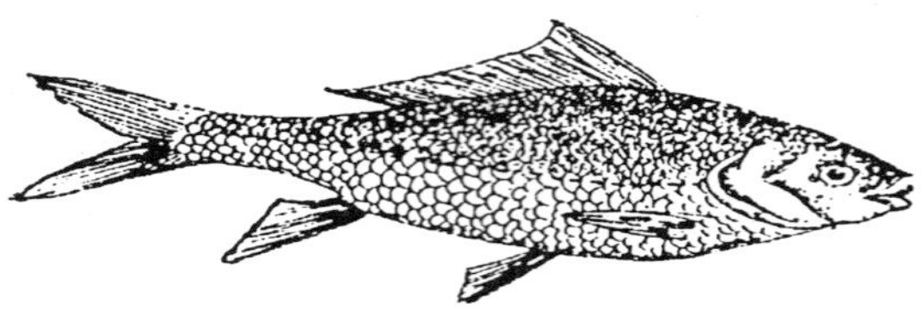

Fig. 1.3 Catla catla

4. *Ctenopharyngodon idellus*

This fish is also known as grass carp or white Amur. The body is elongated and compressed posteriorly. Head is depressed and flattened. Mouth is terminal and lips are thin. The upper jaw is slightly longer than the lower jaw. Eyes are large and lateral in position. Barbels are absent. Dorsal fin is inserted slightly ahead of pelvis. It has ten rays, out of which, seven are branched and three are simple. Anal fin is short with eight branched and two simple rays. Caudal fin is forked. Scales are cycloid. Lateral line is continuous with 40-42 scales.

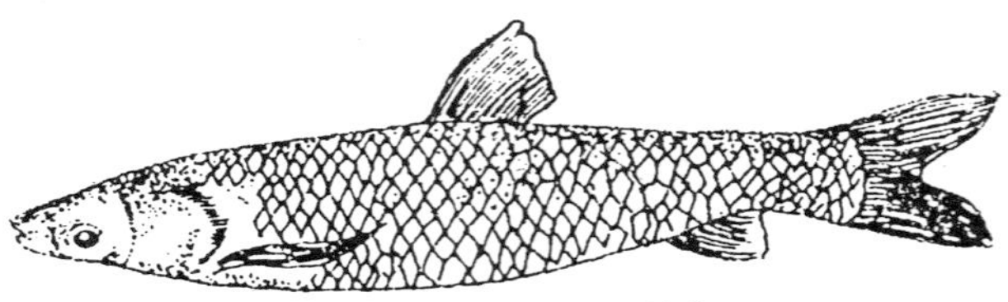

Fig. 1.4 Ctenopharyngodon idellus

5. *Cyprinus carpio*

Head small with protrusible inferior mouth. There are two pairs of barbles. Dorsal fin has three osseus, strong, posteriorly serrated spines. Body oblong and moderately compressed. Abdomen not keeled. As the scales are of golden tinge, the fish assumes gold colour. Fins are pink. There are several varieties of common carp. But three varieties are in India, viz., Scale carp, Mirror carp and Leather carp, of which scale carp is the most widely available and leather carp the least.

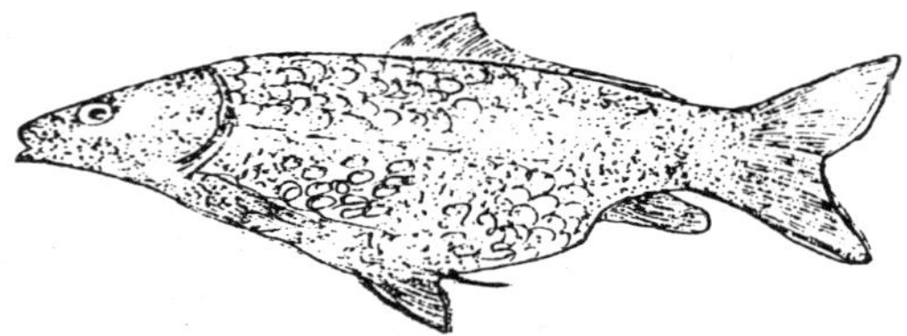

Fig. 1.5 Cyprinus carpio

6. *Labeo rohita*

Body elongated, rather roundish. Head is small. Mouth terminal. Lips thick and fringed. Dorsal profile more arched than ventral profile. Barbels-a-short and thin maxillary pair. Rostral pair rarely present. Dorsal fin originates midway between snout and caudal base. Found in fresh waters.

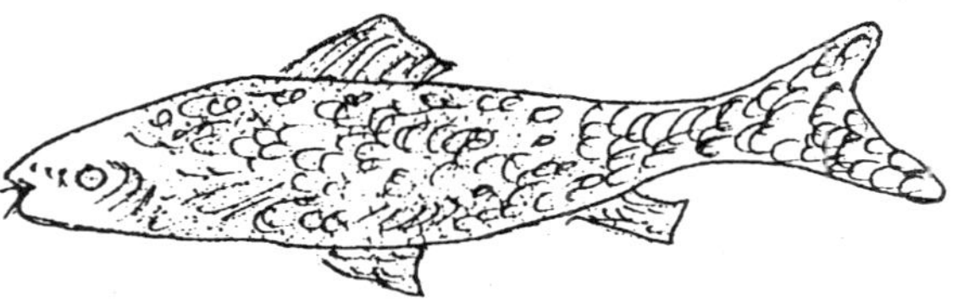

Fig. 1.6 Labeo rohita

7. *Wallago attu*

Body is elongated and compressed and the head is depressed. Mouth subterminal and oblique. Dorsal fin is inserted above half of the pectoral fin. There no adipose dorsal fin in this case. Pectoral fin has caudal fin. Caudal fin is forked is and its lobes are rounded or sometimes pointed. The upper lobe of the caudal fin is longer. Cardiform teeth are present on both the jaws. The palatines are devoid of teeth. Lateral line is complete and simple. Found in freshwaters.

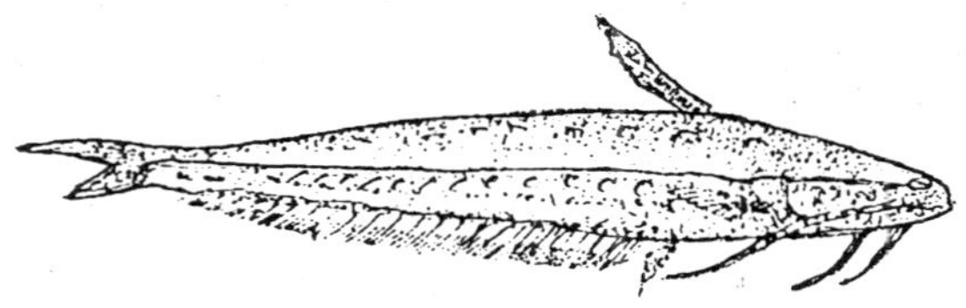

Fig. 1.7 Wallago attu

8. *Channa punctata, C. marulia and C. striata*

They are commonly known as "Snake heads". These are elongated scaly fishes, the head having large shield like scales above. Abdomen is rounded and the head is slightly depressed. Mouth is large and protractile extending below orbit sometimes. Dorsal and anal fins are elongated but these are spineless. The caudal fin is separate and round in shape. Eyes are lateral and moderate in size. Jaws are equal. Teeth are present on the jaws and palate. Accessory respiratory organs are present in the form of folded linings in the paired cavities on the roof of the pharynx.

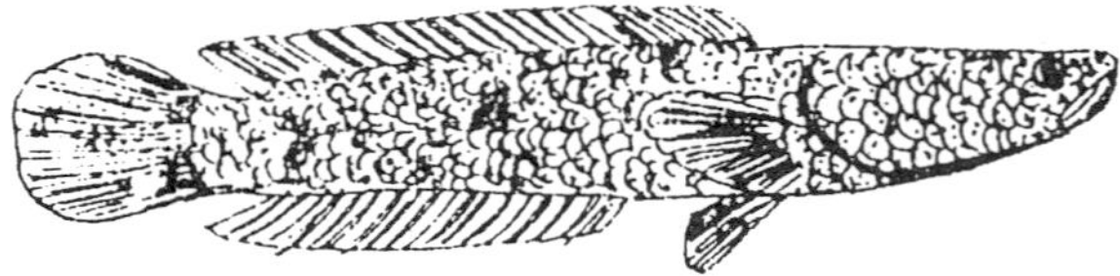

Fig. 1.8 Channa punctata

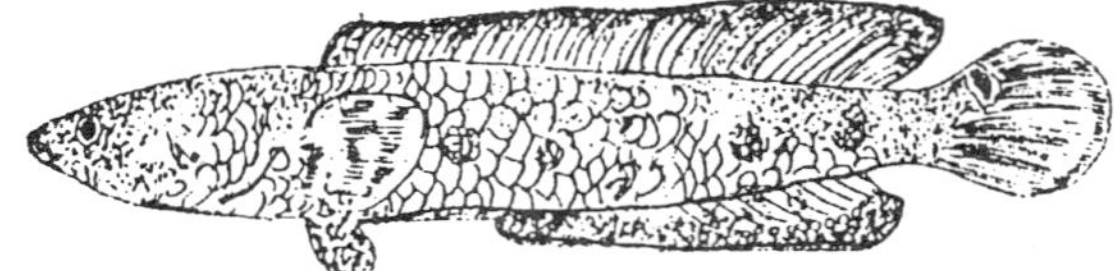

Fig. 1.9 Channa marulia

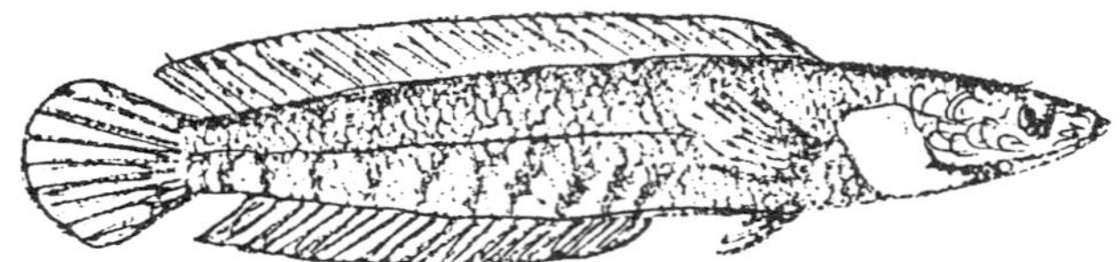

Fig. 1.10 Channa striata

9. *Tilapia mossambica*

T. mossambica is an exotic species introduced in India in as early as 1952. It has a short, compressed and elongated body. The abdomen is rounded. The upper profile of the body is concave. Mouth is terminal and very wide. The snout is rounded and the jaws are equal. Teeth are present in two series, the outer teeth are bicuspid and the others are tricuspid. Dorsal fin is inserted above the base of the pectoral fin and has 15 or 16 spines and 10-11 rays. Anal fin has 3 or 4 spines. Caudal fin is rounded. Scales are cycloid in this fish. Lateral line is incomplete. Upper one has 18-21 scales and the lower one has 10-15 scales only. This fish is not suitable for culture along with the Indian major carps because of its depredation on fry. This fish starts breeding at the age of only two months and it breeds four to six times in a year. Its survival rate is very high, rapidly increase in number.

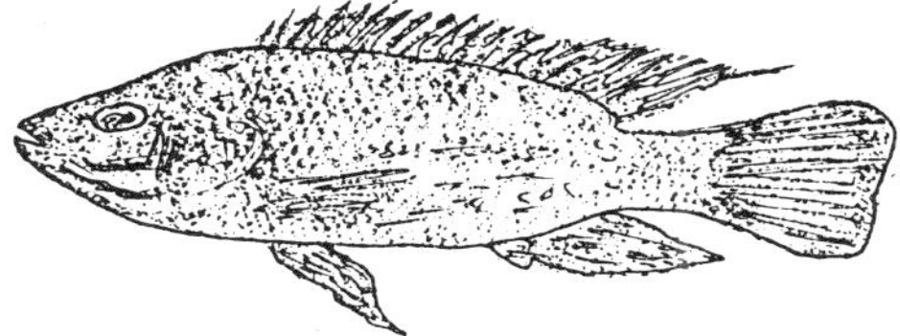

Fig. 1.11 Tilapia mossambica

10. *Clarias batrachus*

Head is 5.5 times in total length of the body. There are two depressions on head, one of it situated behind the eyes and the other towards nare. There are four pairs of barbles. Maxillary pair of barbles reach middle or base of the pectoral fin. Pectoral spine finely serrated. Dorsal fin is longer than anal fin and spineless. Caudal fin is free and rounded. No adipose fin. It is found in fresh and brackish waters.

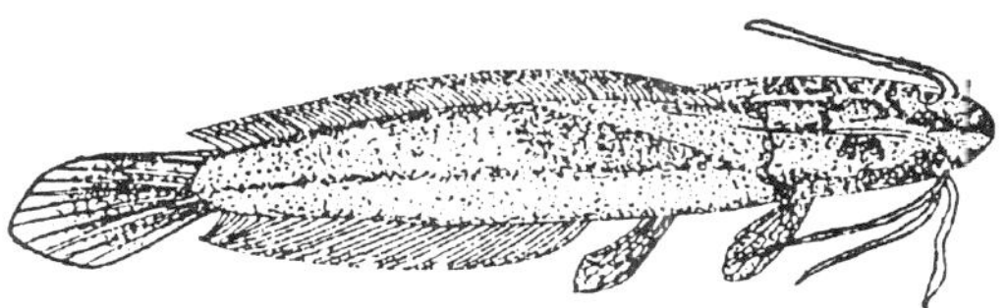

Fig. 1.12 Clarias batrachus

11. *Mugil cephalus*

The head and body are covered with large scales. There are two dorsal fins. Snout rounded pectoral high, no socketted teeth in mouth. Found in coastal waters, migrating into estuaries and rivers.

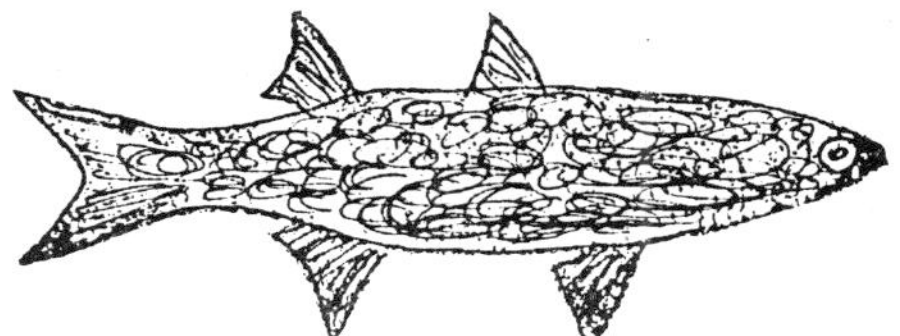

Fig. 1.13 Mugil cephalus

12. *Anabas testudineus*

It possesses well developed accessory air breathing structure and is capable of utilising oxygen present in atmospheric air. It has a laterally flattened body with a prominent triangular head. The mouth is broad. The outer margin of the operculum is provided with spines which aid in locomotion on land. The fins are well developed. The dorsal fin is composed of two portions, the anterior part is larger and supported by stiff spinous fin rays while the posterior part is supported by soft fin rays. The ventral fin is divided into two parts exactly like the dorsal fin. The tail fin is round and numerous fin rays support the fin. The pectoral fins are located near the opecular openings. The basal, supporting the fin is narrow while the terminal end is broad and semicircular. The pelvic fins are placed midventrally close to the pectoral fin. Lower jaw is slightly longer and gill covers serrated. Colour of the body is light to dark green above, greenish yellow to orange below.

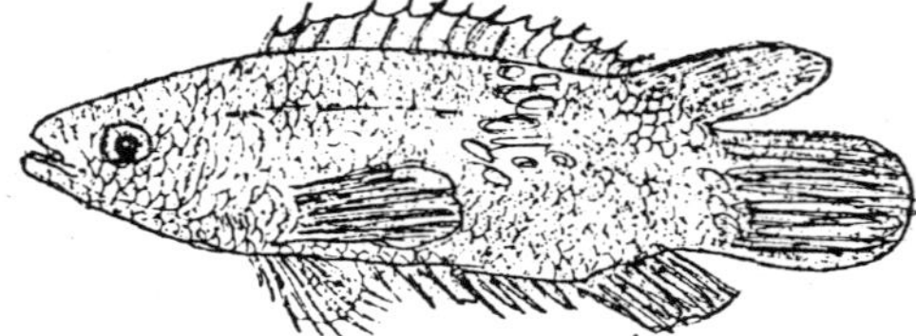

Fig. 1.14 Anabas testudineus

13. *Osphronemus goromy*

Length of the fish is three times more than depth. Four eyes are present on the head. Several rows of small conical teeth are present in jaws. No teeth on paltes. Dorsal fin is shorter than the anal fin.

Spnious of dorsal fin is longer than the soft part. Outer ray of pelvic fin is becoming lighter below. It is an exotic species introduced in India. It is found in fresh and brackish waters.

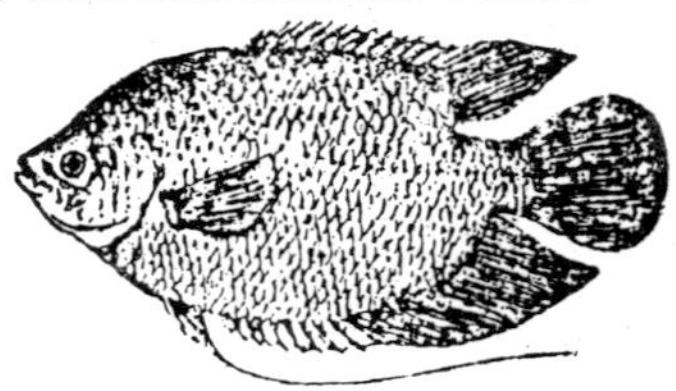

Fig. 1.15 Osphronemus goramy

14. *Etroplus surantensis*

Body laterally compressed, head small, dorsal fin elongated, operculum smooth and anal spines 12-16. Most of the scales above lateral line have a central pearl white spot and on abdomen irregular black spots are present. Fins lead coloured except pectoral which are yellow but jet black at the base. Highly adapted to brackish water and acclimatised to freshwater.

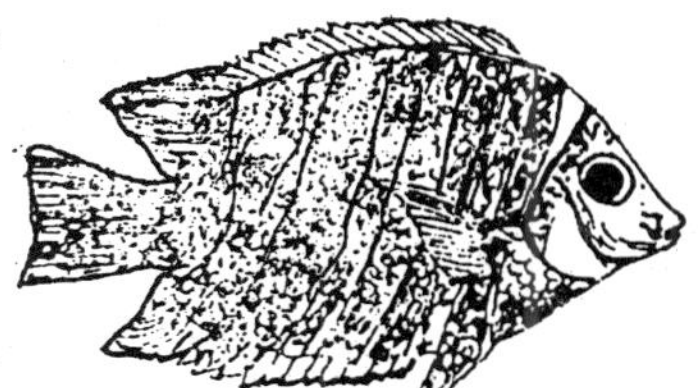

Fig. 1.16 Etroplus suratensis

TAXONOMY OF CULTIVABLE PRAWNS

The prawns belong to the class Crustacea under the phylum Arthropoda; they come under the order Decapoda, suborder Macrura, and tribe Natantia. Their economic importance is pretty high, as they are among the choicest food items under marine food resources. Apart from the internal demand they support a very valuable trade for export which since the early fifties of this century with the introduction of sophisticated processing techniques of freezing and canning, caught the foreign markets in the U.S.A. some of the European countries and Japan. India's prawn grounds are among the worlds most productive ones noted both for their high yields and superior quality of the species occurring in the fairly high proportions.

The prawns are broadly divided into penaeid and non-penaeid varieties. The penaeid prawns (usually called as shrimps) under the family Penaeidae are very important, for they grow to a large size suitable for export trade and are obtained in appreciable quantities. The non-penaeid prawns come under several families, viz., Hippolytidae, Palaemonidae, Pandalidae and Sergistidae.

Phylum	Arthropoda
Class	Crustacea
Sub Class	Malacostraca
Order	Dacapoda

Sub order	Macrura	
Tribe	Natantia	
Family	Palaemonidae	*Macrobrachium rosenbergii*
		Macrobrachium malcomsonii
Family	Penaeidae	*Penaeus monodon*
		Penaeus indicus
		Penaeus semisulcatus
		Metapenaeus monoceros
		Metapenaeus dobsonii
		Metapenaeus brevicornis

1. *Macrobrachium rosenbergii*

This species is suitable for culture in confined water. The rostrum is longer, sword shaped and bears 13-14 rostral spines dorsally and 11 rostral spines ventrally. The second pair of walking legs of males develop abnormally with well developed chela and thus sexual dimorphism is exhibited. The legs are stout. It ocurs in rivers, estuaries and coastal areas. It migrates to estuaries during breeding season.

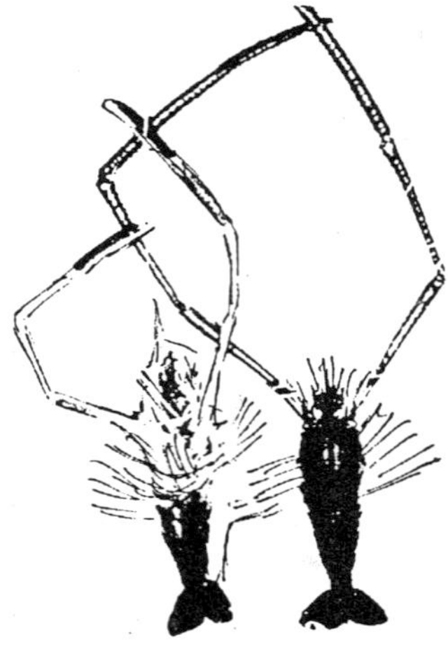

Fig. 1.17 Macrobrachium rosenbergii

2. *Macrobrachium malcomsonii*

The rostrum is short and straight. It bears 11-12 dorsal spines and 4-5 ventral spines. The second pair of walking legs of males develop chela and thus sexual dimorphism is exhibited. The second pair of legs are longer than the body and are stout. Migrate from rivers to estuaries during breeding season.

3. *Penaeus monodon*

Smooth carapace. Rostrum strongly sigmoidal. Rostral formula 7-8 by 2-3, abrostral carina and groove narrow not reaching beyond the epigastric spine (first spine) on the carapace. Carapace with well defined antennal and hepatic spines. The hepatic carina is horizontally straight. Body with transverse bands. The colour of live prawn often depends on the area of collection. Generally dark brown with emphasis on black or reddish in clear deep waters. Antennae dark and light colour. Pleopods are brown to blue with distinct yellow reddish tinge consists of two lateral lobes closely meeting along the median line.

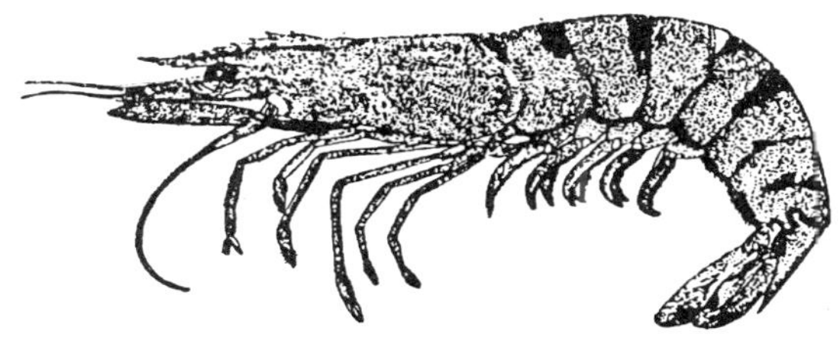

Fig. 1.18 Penaeus monodon

4. *Penaeus indicus*

This species shows large variations in rostral length, in young forms. In adults the length is reduced and thickened at the base. It has a clear double curve with 7-9 by 4-6 teeth. The post rostral crest is not markedly elevated but is triangular in profile and faintly canaliculate. The gastro-orbital carina is well defined. No bands on the body. The colour is creamy white with dispersed light pigmentation.

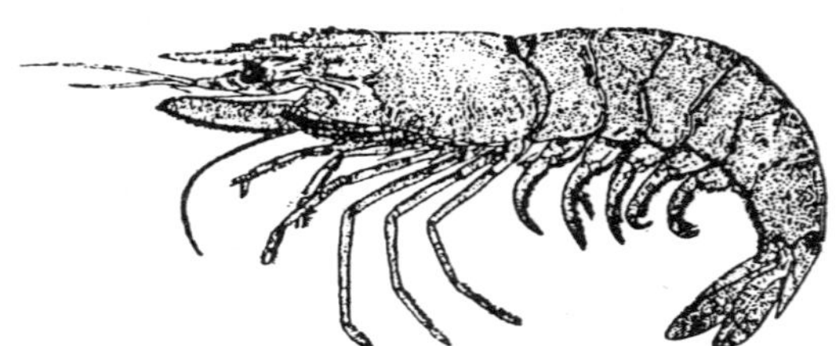

Fig. 1.19 Penaeus indicus

5. *Metapenaeus monoceros*

Body covered with short tormentum, rostrum nearly straight giving a knife like appearance reaching nearly to or a little beyond, the tip of antennular peduncle. Rostral formula 9-12. The 2nd to 6th abdominal terga, usually the last are carinated mid dorsally, the last 3 strongly so. In adult males the 5th periopad has the proximal and of merus notched on its outer side with a spine. No exopodite on the 5th pereiopod. Petasma with large distal lobes distally with no lateral projections. Body gives a strongly speckled appearance due to light pigmentation. General colour emphasis varies from grey to green.

Fig. 1.20 Metapenaeus monoceros

6. *Metapenaeus dobsoni*

Body tomentose in patches. Rostrum double curved with 8-9 dorsal teeth and no teeth. Spine on basis of their pair of walking legs in male long. No exopod on the fifth pair of walking legs. Reddish-brown spots scattered on semi transparant body. Rostrum, carapace and pleural edges deep brown; antennules and antennae dotted red. Double rows of reddish spots on telson.

Fig. 1.21 Metapenaeus dobsoni

2

COMMON FISH OF INDIA

In this chapter, some very common fishes of India have been described with their latest classification, diagnostic features and distribution, in addition to the once described in chapter 1.

1. *Channa punctatas*

Channa marulias

Channa striatas

Phylum	:	Chordata
Sub phylum	:	Vertebrata
Class	:	Pisces
Sub class	:	Teleostomi
Super order	:	Acanthopterygii
Order	:	Channiformes
Family	:	Channidae

They are commonly known as "Snake heads". These are elongated scaly fishes, the head having large shield like scales above. Abdomen is rounded and the head is slightly depressed. Mouth is large and protractible extending below orbit sometimes. Dorsal and anal fins are elongated but these are spineless. The caudal fin is separate and round in shape. Eyes are lateral and moderate in size. Jaws are equal. Teeth are present on jaws and palate.

Channa punctatus is present in muddy streams and is generally greenish-brown above and yellowish below in colour. But in brackish waters they exhibit purple in colour. A dark variety with black spots are also present. In this species, pelvic fins has more than half length of pectoral fin. 15-16 predorsal scales also are present. These are carnivorous.

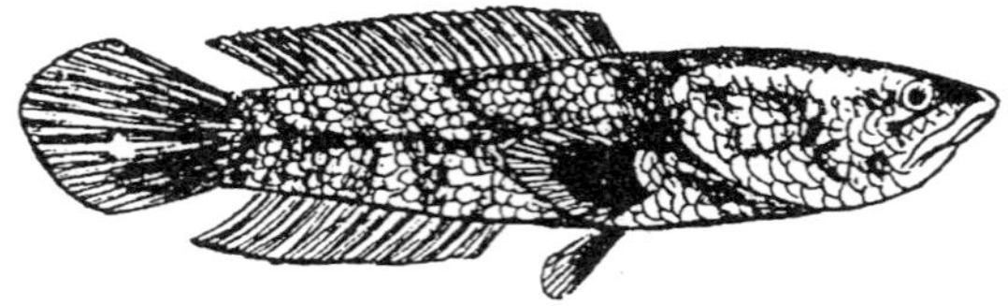

Fig. 2.1 Channa punctatus

Channa marulius can be grown in wells, and generally feed on kitchen refuse, frog and dead animals. On the basal portion of the caudal fin a white-edged ocellus is present. The pelvic fin is less than half in length of pectoral fin. Pectoral fin have darker and lighter patches at certain zones.

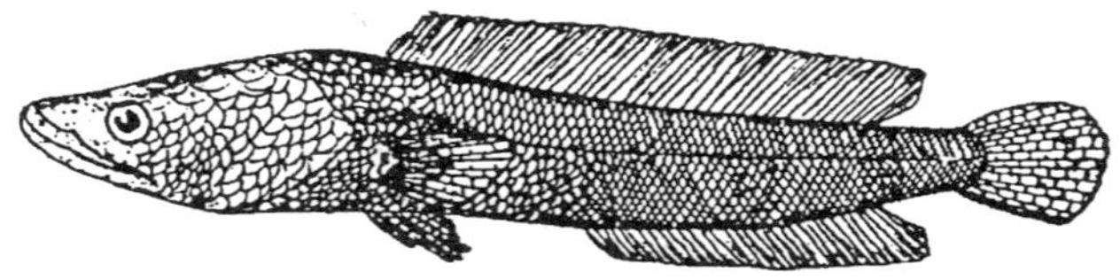

Fig. 2.2 Channa marulius

Channa striatus has dark-brown or black colour in the upper part of the body but the lower part the colour is orange or yellowish. They are generally carnivorous in their feeding habit. 18-29 predorsal scales are present. The length is 30-35 cm.

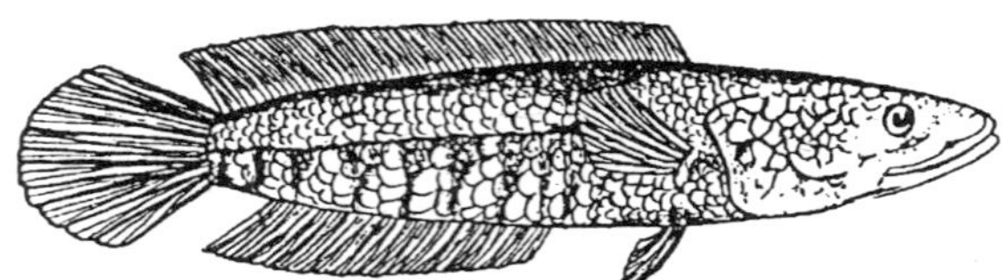

Fig. 2.3 Channa striatus

All these species are distributed throughout India, Nepal, Bangladesh, China, Pakistan, Sri Lanka and some other places.

2. *Gambusia affinis patruelis*

Phylum	:	Chordata
Sub phylum	:	Vertebrata
Class	:	Pisces
Sub class	:	Teleostomi
Super order	:	Atherinomorpha
Order	:	Atheriniformes
Family	:	Poecilidae

Gambusia is a larvivorous exotic fish with cylindrical and compressed body. The fish abdomen is rounded and the head is short and the snout is pointed. The mouth is small and oblique. The eyes are present in the centre of the head. Upper jaw is short and the lower jaw is slightly turned upwards. Teeth are absent. In case of male fish the dorsal fin is present in the middle of the body. In female fish, dorsal fin is present midway between the front margin of the eye and tip of the caudal fin. The dorsal fin has 6-12 rays. The anal fin in male has a longer process which distinguishes it from the female. This fish is known as mosquito fish because of its liking for mosquito larvae. This fish is majorly found in Asia.

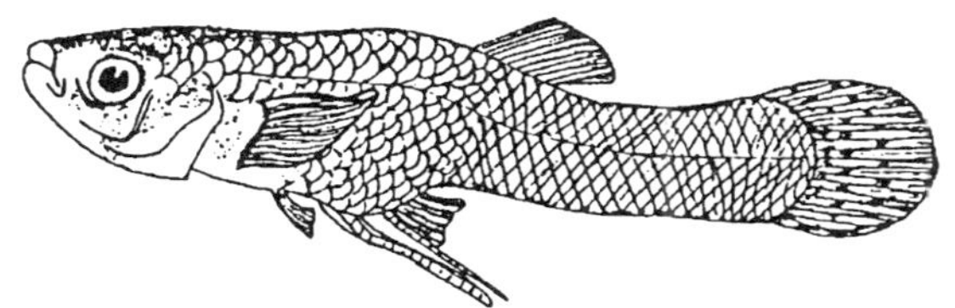

Fig. 2.4 Gambusia affinis male

3. *Lebistes reticulatus*

Phylum	:	Chordata
Sub phylum	:	Vertebrata
Class	:	Pisces
Sub class	:	Teleostomi
Super order	:	Atherinomorpha
Order	:	Atheriniformes
Family	:	Poecilidae

Eyes are big and situated on the dorsal profile of the head. Caudal fin is elongated. Black spots or some other briliant coloured patches are present on the body. Functional hermaphroditesm has been reported in *Lebistes*. It is also a larvivorous fish like *Gambusia.*

Fig. 2.5 Lebistes reticulatus male

4. *Xenentodon cancila*

Phylum	:	Chordata
Sub phylum	:	Vertebrata
Class	:	Pisces
Sub class	:	Teleostomi
Super order	:	Atherinomorpha
Order	:	Atheriniformes
Family	:	Belonidae

This fish body is sub-cylindrical, stout and compressed. The abdomen of this fish is rounded and the head is pointed. Snout is pointed. Eyes are moderate but superior in position. The upper part of the body is distinct by the presence of black dots. Silvery streak with dark margin extends along the body from orbit to middle of caudal base. Along the upper surface of the head, there is a deep longitudinal groove. Jaws are provided with sharp teeth. The lower jaw is slightly larger in size. Dorsal fin has 15-18 rays. Anal fin has 15-19 rays. Caudal fin is truncate. The scales are irregularly arranged.

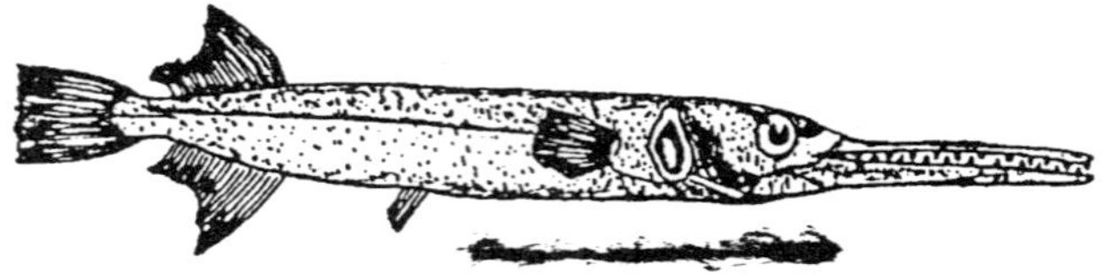

Fig. 2.6 Xenentodon cancila

5. *Notopterus chitala*

Phylum	:	Chordata
Sub phylum	:	Vertebrata
Class	:	Pisces
Sub class	:	Teleostomi
Super order	:	Osteoglossomorpha
Family	:	Notopteridae

Notopterus body is oblong, laterally compressed and deep and it is covered with overlaping scales. The minute scales are devoid of photophores. The dorsal fin is not forked. Eyes are moderate in size. Teeth are present on premaxillae, maxilla, vomer, palatine and pterygoid. Tongue is also provided with teeth. Snout is prominent. The body is brown or greenish along narrow back having 15 silvery transverse bars meeting across dorsal ridge.

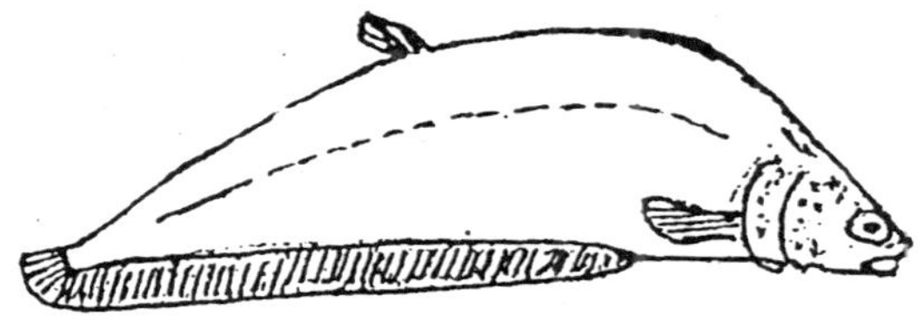

Fig. 2.7 Notoptelus chitala

Fig. 2.8 Notopterus notopterus

6. *Tetraodon cutcutia*

Tetraodon fluviatilis

Phylum	:	Chordata
Sub phylum	:	Vertebrata
Class	:	Pisces
Sub class	:	Teleostomi
Super order	:	Acanthopterygii

Order	:	Tetraodontiformes
Family	:	Tetraodontidae

The body is short and the back is broad. The head is oval, snout blunt, and the abdomen is rounded. Mouth is terminally located and transverse. Lips are very thick and fleshy. Both the jaws have a median surface. The ventral fin is absent. The air bladder is horse-shoe shaped. *T. fluviatilis* body is cylindrical and elongated. Lateral line is absent. The head is dark-grey in colour.

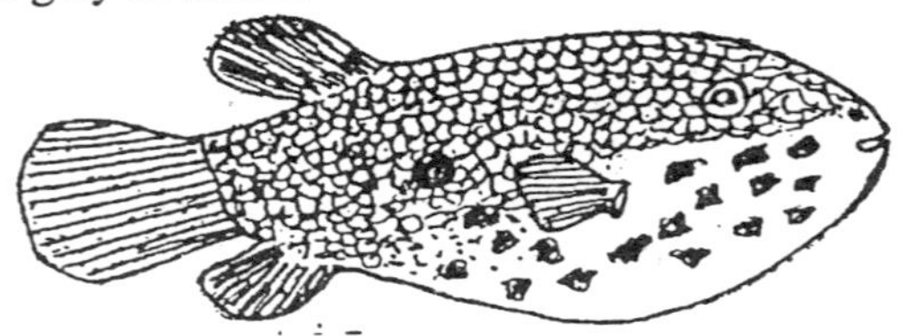

Fig. 2.9 Tetraodon cutcutia

Fig. 2.10 Tetraodon fluviatilis

7. *Anabas testudineus*

Phylum	:	Chordata
Sub phylum	:	Vertebrata
Class	:	Pisces
Sub class	:	Teleostomi
Super order	:	Acanthopterygii
Order	:	Perciformes
Family	:	Anabantidae

Anabas body is stout, oblong, almost cigar shaped body and compressed. The head colour is greenish brown. Snout is slightly conical. The jaws are equal and lips are thin. Single dorsal fin covering the entire length of the trunk. Caudal fin is rounded. Ctenoid scales are present. Teeth are absent. Due to the presence of accessory respiratory organ and it can survive in moist air for about one week. Small teeth are present on the vomers. Eyes are lateral in position and are large.

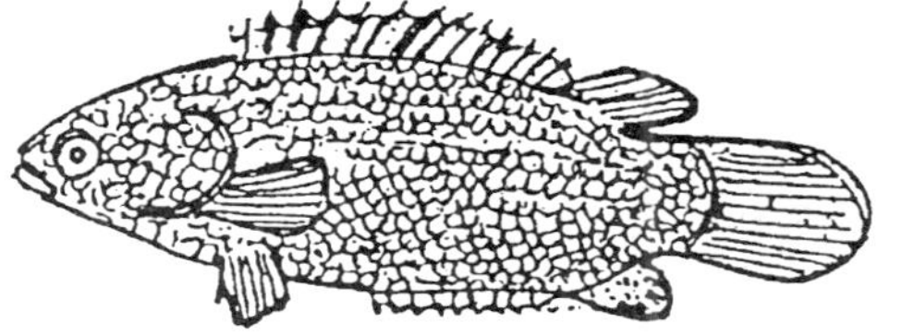

Fig. 2.11 Anabas testudineus

8. *Chanda nama* and *Chanda ranga*

Phylum	:	Chordata
Sub phylum	:	Vertebrata
Class	:	Pisces
Sub class	:	Teleostomi
Super order	:	Acanthopterygii
Order	:	Perciformes
Family	:	Chanididae

Body is short, deep and compressed. Head is short and compressed. Snout is sharp. Villiform teeth are present on the jaws, palate and tongue. Two dorsal fins are present, which are continuous. Caudal fin is forked. Cycloid scales are present. Eyes are large and superior.

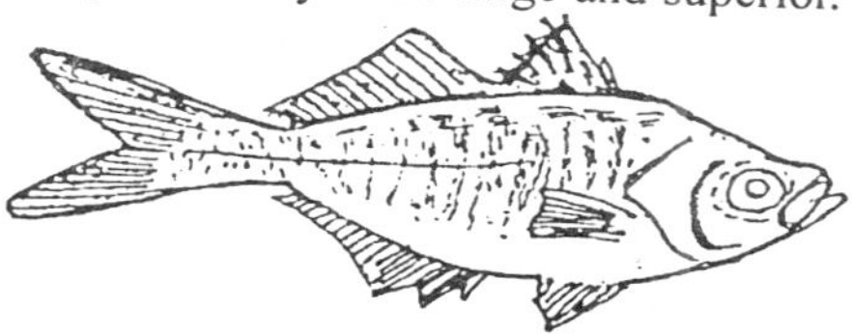

Fig. 2.12 Chanda nama

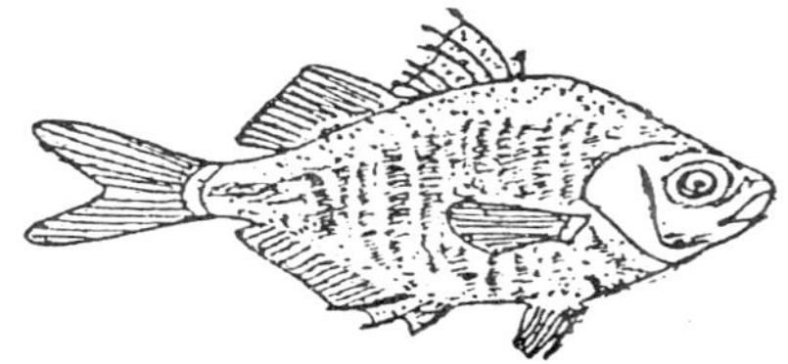

Fig. 2.13 C. ranga

9. *Tilapia mossambica*

Phylum : Chordata

Sub phylum	:	Vertebrata
Class	:	Pisces
Sub class	:	Teleostomi
Super order	:	Acanthopterygii
Order	:	Perciformes
Family	:	Cichlidae

Tilapia is an exotic species. Body is short, compressed and elongated. The abdomen is rounded. Mouth is terminally located and wide. The jaws are equal and the snout is rounded. Teeth are present in two series, the outer teeth are bicuspid and the others are tricuspid. Dorsal fin has 15-16 rays. Caudal fin has 10-11 rays. This fish is not compatible for culture with Indian Major carps, because it will feed on fry of carps. *Tilapia* starts breeding at the age of two months and breeds 4 to 6 times in a year. *Tilapia* is a very hard species, having high survival rate and rapidly increase in number.

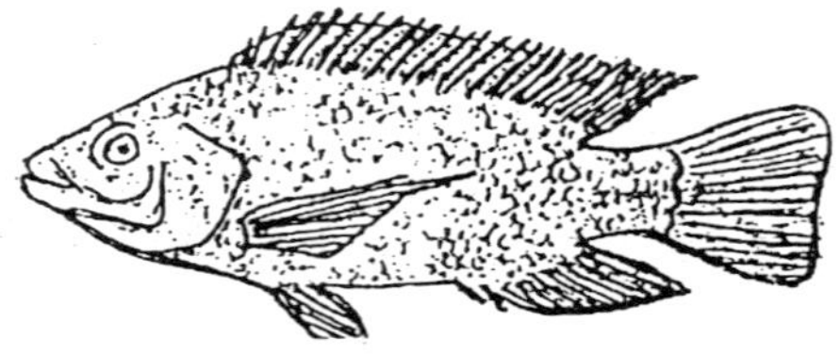

Fig. 2.14 Tilapia mossambica

10. *Clarias batrachus*

Phylum	:	Chordata
Sub phylum	:	Vertebrata
Class	:	Pisces
Sub class	:	Teleostomi
Super order	:	Ostariophysi
Order	:	Siluriformes
Family	:	Claridae

Body of this fish is elongated and compressed with rounded abdomen. Head is depressed. Upper jaw is longer than the lower one. Villiform teeth are present on jaws and palate. Four pairs of barbels are present—two pairs of mandibular, one pair of maxillary and nasal each. The eye size is small, because of its adaptation to cope up with muddy habitat. Dorsal fin is very long. Caudal fin is rounded and free. Lateral line is complete.

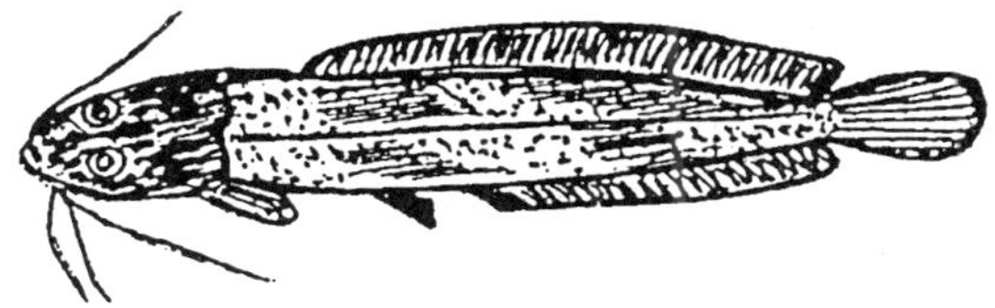

Fig. 2.15 Clarias batrachus

11. *Glyptothorax saisii*

Phylum	:	Chordata
Sub phylum	:	Vertebrata
Class	:	Pisces
Sub class	:	Teleostomi
Super order	:	Ostariophysi
Order	:	Siluriformes
Family	:	Sisoridae

Glyptothorax body is highly granulated, elongated and moderately depressed. Abdomen may be flat or slightly rounded. Mouth is inferior in its position and is very narrow. Head is small. Snout is conical in its shape. Eyes are dorsal in its position and are small. Lips are thick and papillated. Upper jaw is longer than the lower jaw. Villiform teeth present on the jaws and palate. Ventral surface of the body in between the pectoral fins is highly modified and provided with adhesive apparatus, a hill-stream adaptation. Four pairs of barbels are present, i.e., one pair of nasal, one pair of maxillary and two pairs of mandibular. The maxillary barbels are provided with broad bases. Pelvic fin has 6 rays. Anal fin is short and it has 7-14 rays. Caudal fin is deeply forked. Air bladder also present.

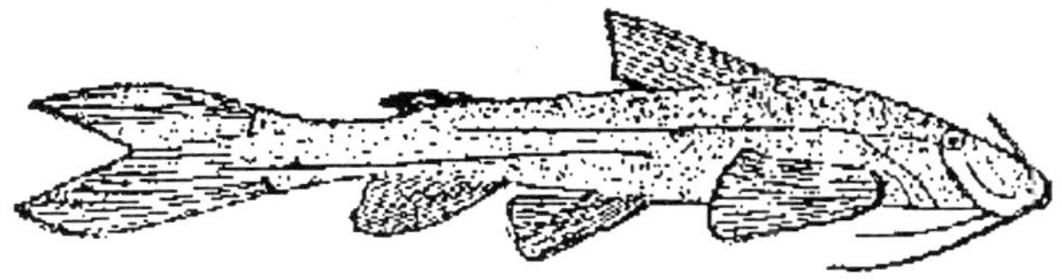

Fig. 2.16 Glyptothorax saisii

12. *Mystus cavasius*

Mystus tengra, ***Mystus vittatus*** and ***Mystus or***

Phylum	:	Chordata
Sub phylum	:	Vertebrata
Class	:	Pisces
Sub class	:	Teleostomi
Super order	:	Ostariophysi
Order	:	Siluriformes
Family	:	Bagridae

Fish body and head are compressed. Body is moderately elongated. Snout is rounded. Mouth is terminally placed. Eyes are relatively small. Villiform teeth present on the palate and on upper jaw. There are 4 pairs of barbels—one pair of maxillary, one pair of nasal and two pairs of mandibular. Anal fin is short. Caudal fin is forked. Lateral line is present. In *M. cavasius,* maxillary barbels reach the base of the caudal fin. At the base of the dorsal fin, a dark spot is present. In *M. tengra*, maxillary barbels reach the base of the pelvic fins. The maxillary barbel reaches the base of the ventral fin. The external mandibular barbel reaches the middle of the pectoral fin and the internal barbel reaches only up to the base of the pectoral fin. Mandibular barbels are white with a black streak. In *M. vittatus*, a dark colour spot is present. At the base of the caudal fin there is no spot. Caudal fin is forked and the upper lobe is ionger than the lower one. In the case of *M. aor*, the maxillary barbels are exceptionally long. Snout is very broad.

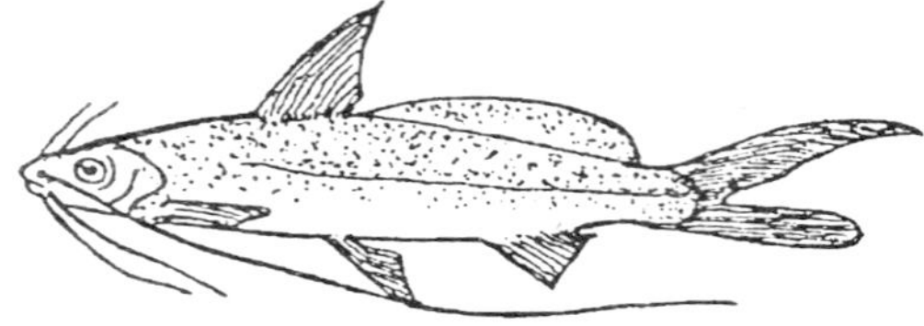

Fig. 2.17 Mystus cavasius

Fig. 2.18 M. Tengra

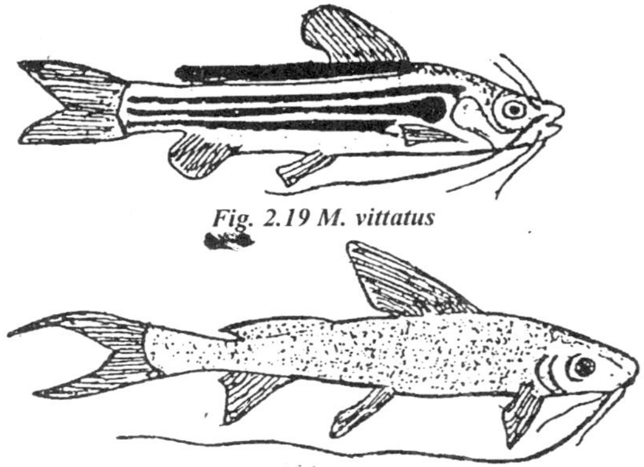

Fig. 2.19 M. vittatus

Fig. 2.20 M. aor

13. Wallago attu

Phylum	:	Chordata
Sub phylum	:	Vertebrata
Class	:	Pisces
Sub class	:	Teleostomi
Super order	:	Ostariophysi
Order	:	Siluriformes
Family	·	Silurudae

Wallago's body is elongated and compressed. Mouth is subterminal and oblique. Jaws are unequal—lower jaw being longer than the upper one. Villiform teeth are present in bands on jaws and in patches on palate. 2 pairs of barbels are present—one pair maxillary and one pair mandibular. Maxillary barbels are longer than the mandibular barbels. Adipose dorsal fin is absent. Caudal fin is forked and its lobes are rounded. The upper lobe of the caudal fin is longer. Lateral line is complete and simple.

Fig. 2.21 Wallago attu

14. *Tor tor*

Phylum	:	Chordata
Sub phylum	:	Vertebrata
Class	:	Pisces
Sub class	:	Teleostomi
Super order	:	Ostariophysi
Order	:	Cypriniformes
Family	:	Cyprinidae
Sub Famil	:	Cyprininae

This fish body is generally compressed and elongated. The ventral profile is more arched. The abdomen is round. Eyes are relatively large, and their size varies accordingly to age. The caudal fin is deeply forked. Scales are large and the lateral line is complete. Generally fins are reddish yellow in colour.

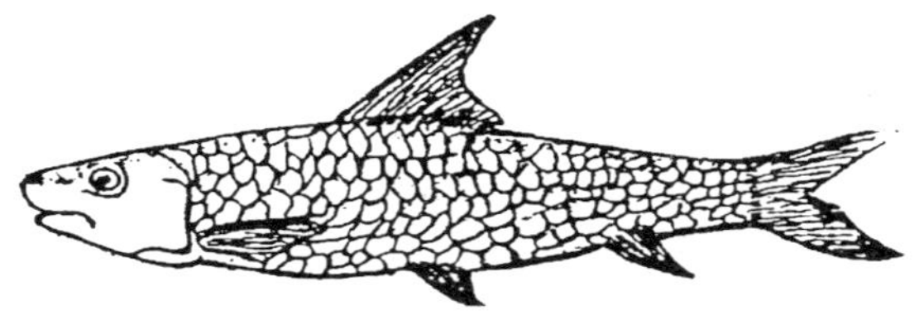

Fig. 2.22 Tor tor

15. *Syngnathus argyrostictus*

Phylum	:	Chordata
Sub phylum	:	Vertebrata
Class	:	Pisces
Sub class	:	Teleostomi
Super order	:	Acanthoptergii
Family	:	Syngnathidae

The body of this fish is normally elongated and has distinct ridges. The abdomen is rounded. There is no spiny dorsal fin. Ventral and caudal fins are absent. Anal fin is very much reduced with one or two rays. In male fish, brood pouch is present, which has longitudinal opening. The snout is cylindrical. Jaws are transformed to produce a beak. There is no barbel. Lateral line is complete. Snout is longer than remaining part of the head. The trunk is provided with seven longitudinal series of pearly ocelli.

Fig. 2.23 Syngnathus

16. *Ctenopharyngodon idella*

Phylum	:	Chordata
Sub phylum	:	Vertebrata
Class	:	Pisces
Sub class	:	Teleostomi
Super order	:	Ostariophysi
Order	:	Cypriniformes
Family	:	Cyprinidae
Sub Family	:	Cyprininae

This fish is commonly known as grass carp or white Amur. This fish body is compressed. Head is flattened. Mouth is terminally located and the lips are very thin. The upper jaw is longer than the lower jaw. Eyes are lateral in position and are relatively large. Barbels are absent. Dorsal fin has 10 rays. Out of which 7 are branched and 3 are simple. Anal fin has 8 branched and 2 simple rays. Caudal fin is forked. Cycloid scales are present. Lateral line is continuous with 40-42 scales.

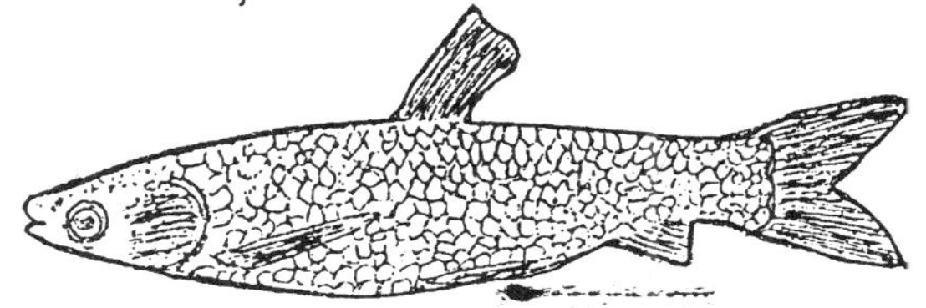

Fig. 2.24 Ctenopharyngodon idella

16. *Lates calcarifer*

Phylum	:	Chordata
Sub phylum	:	Vertebrata

Class	:	Pisces
Sub class	:	Teleostomi
Super order	:	Acanthopterygii
Order	:	Perciformes
Family	:	Centropomidae

This fish has deep and compressed body with moderate sized head, which also compressed. Mouth is very wide. Eyes are moderate in its size and are supero-lateral in their position. Lower jaw is longer than the upper jaw. Villiform teeth are present on the palate. Tongue is devoid of teeth. Two dorsal fins are continuous and they are united at the base. The first dorsal fin has 7-8 spines, but the second has 11-13 rays. Anal fin has 3 spines and 8-9 rays. Caudal fin is rounded. Lateral line is sligthly curved and possess 52-60 scales. Ctenoid type of scales are present.

Fig. 2.25 Lates calcarifer

17. *Monopterus cuchia*

Phylum	:	Chordata
Sub phylum	:	Vertebrata
Class	:	Pisces
Sub class	:	Teleostomi
Super order	:	Acanthopterygii
Order	:	Synbranchiformes
Family	:	Synbranchiadae

The fish body is elongated, and eel-shaped. The gill opening is present in the form of a slit. Wide mouth. Carnivorous in feeding habit. Dorsal fin is reduced. The skin is glandular. Premaxillae, palatine and lower jaw are provided with teeth. The gill openings are triangular without lateral folds. Well developed lateral line is present. Scales are minute and are arranged in longitudinal rows.

Fig. 2.6 M. Cuchia

18. *Labeo calbasu*

Labeo rohita and

Labeo gonius

Phylum	:	Chordata
Sub phylum	:	Vertebrata
Class	:	Pisces
Sub class	:	Teleostomi
Super order	:	Ostariophysi
Order	:	Cypriniformes
Family	:	Cyprinidae
Sub Family	:	Cyprininae

The body of all the above fish species are moderately elongated body and rounded abdomen. Mouth is mostly inferior. Jaws are provided with transverse folds. Two or four barbels may be present. Pharyngeal teeth are hooked and are in three rows. Dorsal fin is inserted ahead of pelvic fins. Anal fin is short. Caudal fin is forked. In *L. calbasu* 5-666 scales are present between lateral line and pelvic fin. Dorsal fin has 16018 rays. In *L. gonius*, 9-13 scales are present between lateral line and pelvic fin base. One pair of rostral and one pair of maxillary barbels present. In *L. rohita*, 6 scales present between lateral line and pelvic fin base. The colour of the fish is black along the sides and silvery in abdominal area.

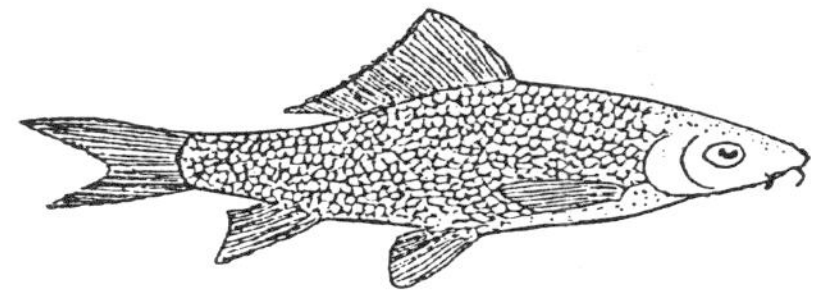

Fig. 2.27 Labeo calbasu

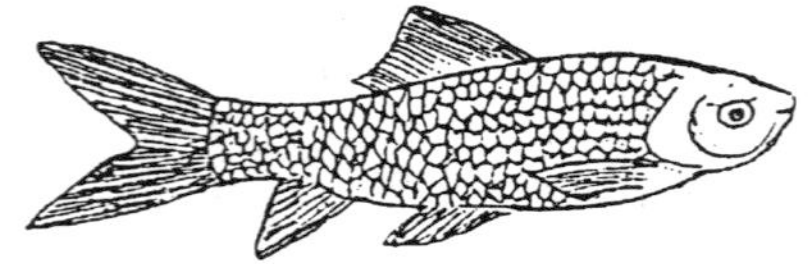

Fig. 2.28 L. rohita

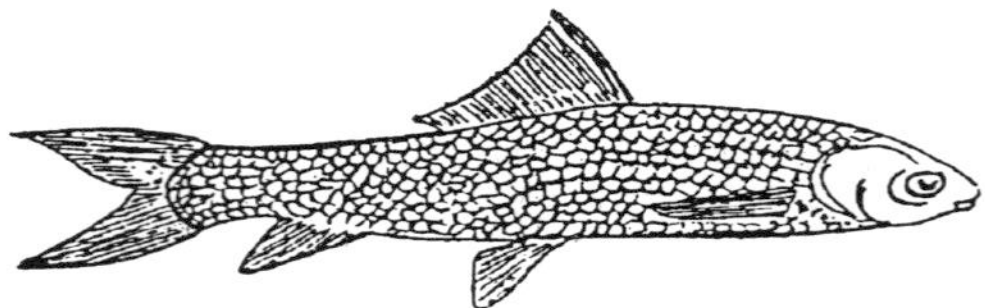

Fig. 2.29 L. gonius

19. *Nandus nandus*

Phylum	:	Chordata
Sub phylum	:	Vertebrata
Class	:	Pisces
Sub class	:	Teleostomi
Super order	:	Acanthopterygii
Order	:	Perciformes
Family	:	Nandidae

The body is compressed and has rounded abdomen. Head is relatively large and compressed. Mouth is terminal in position with wide cleft. Eyes are large and lateral in position. Villiform teeth present on palate and tongue. Dorsal fin is inserted above pectoral base. Ctenoid scales present. Lateral line have 46-48 scales. Three vertical wavy bands present on the side of the body. Opercle has one spine.

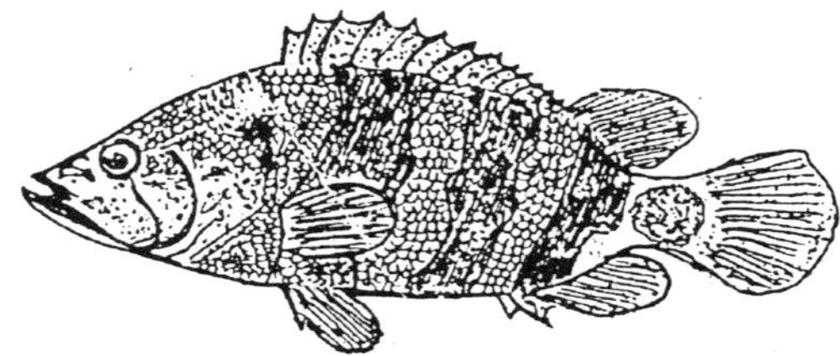

Fig. 2.30 Nandus nandus

20. *Zenarchopterus dispar (Hemiramphus valenciennes)*

Phylum	:	Chordata
Sub phylum	:	Vertebrata
Class	:	Pisces
Sub class	:	Teleostomi
Super order	:	Antherinomorpha
Order	:	Atheriniformes
Family	:	Hemiramphidae

Fish body is elonagted, sub-cylindrical and laterally compressed. The upper jaw is short and triangular in its shape. The lower jaw is prolonged into slender beak bearing sensitive fringe. Pointed head and snout are the characteristic feature. Wide mouth superior in its position. Villiform teeth present on jaws. Dorsal fin has 11-14 rays. Anal fin has 10-12 rays. In male sex, 6th and other rays are modified/transformed. They are much elongated. The anal fin is separated into 1-3 distinct parts. The caudal fin is rounded and never forked. Lateral line is complete and possess 50 scales.

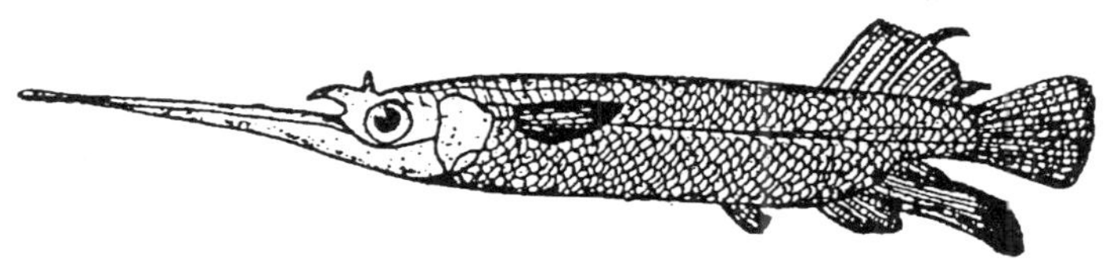

Fig. 2.31 Zenarchopterus dispar

21. *Lepidocephalichthys guntea*

Phylum	:	Chordata
Sub phylum	:	Vertebrata
Class	:	Pisces
Sub class	:	Teleostomi
Super order	:	Ostariophysi
Order	:	Cypriniformes
Family	:	Cobitidae
Sub Family	:	Cobitinae

This body is generally elongated and moderately compressed. Head is short and conical in its shape. Abdomen is rounded. Mouth is slightly arched and inferior in its position. Eyes are relatively small and superior in its position. Two pairs of rostral and one pair of maxillary barbels are present. Dorsal fin is inserted slightly ahead of pelvic fins. In males, the inner ray of pectoral fin is ossified as a flat osseousm plate like structure. Caudal fin is rounded. Numerous rows of dark spots are present on the caudal and dorsal fins. Generally the fish is dirty yellowish brown.

Fig. 2.32 Lepidocephalichthys guntea

22. *Glossogobius giuris*

Phylum	:	Chordata
Sub phylum	:	Vertebrata
Class	:	Pisces
Sub class	:	Teleostomi
Super order	:	Acanthopterygii
Order	:	Perciformes
Family	:	Gobiidae
Sub Family	:	Gobiinae

Body is elongated. Anterior portion of body is cylindrical. Head and snout are pointed. Mouth is slightly oblique. Several rows of villiform teeth are present. Outer and inner rows of teeth enlarged. Bilobate tongue is present. Anterior scales are cycloid and the posterior scales are ctenoid. Anal fin posteriorly pointed. Caudal fin is oblong. Ventral fins are united at their bases forming a cupshaped structure. Two dorsal fins are separated by a short interpace. The dorsal fin is inserted above half or three-fourth of pector

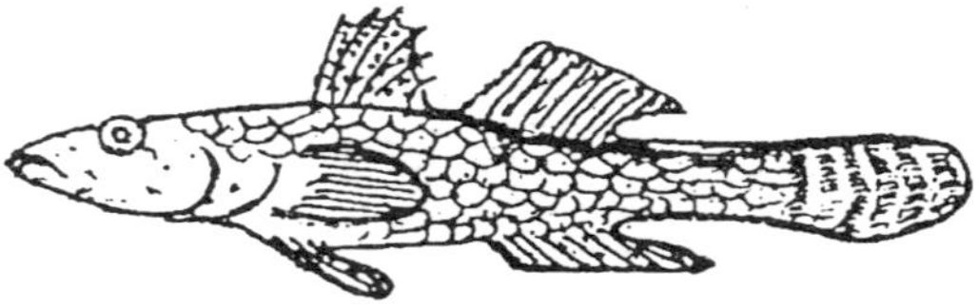

Fig. 2.33 Glossogobius qiuris

23. *Heteropneustes fossilis*

Phylum	:	Chordata
Sub phylum	:	Vertebrata
Class	:	Pisces
Sub class	:	Teleostomi
Super order	:	Ostariophysi
Order	:	Siluriformes
Family	:	Heteropneustidae

Body is elongated. Head size is moderate. Mouth is narrow and terminally located. Unequal jaws present. Villiform teeth present on jaws and palate. There are 8 barbels—2 maxillary, 2 nasal and 4 mandibular. Accessory respiratory organ is present in the form of airsacs which extend into the caudal region. Paired fin are horizontally inserted. Dorsal fin is short and has no spine. Pectoral fin has a strong spine. Caudal fin is rounded and anal fin by a deep notch. Anal fin is very long. The fish is brown in its colour.

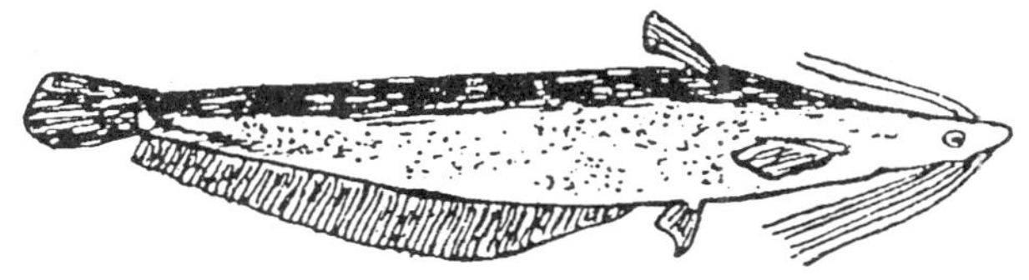

Fig. 2.34 Heteropneustes fossilis

24. *Setipinna phasa*

Phylum	:	Chordata
Sub phylum	:	Vertebrata
Class	:	Pisces
Sub class	:	Teleostomi
Super order	:	Clupeomorpha
Order	:	Clupeoformes
Family	:	Engraulidae
Sub Family	:	Engraulidae

Body is elongated. The abdomen is serrated with 15-23 pre-pelvic and 6-12 post-pelvic scutes. Snout is obtused. Head is short and compressed. Eyes are large and lateral in position. Adipose eyelid is present. Maxilla extends behind the eye. Jaws are unequal and the lower jaw is shorter than the upper one. Small teeth present on the jaws and palate. Dorsal fin is inserted behind the anal fin. The dorsal spine is short but strong. The anal fin is very long. Lateral line is absent. Caudal fin is forked and the lower lobe is larger than the upper one.

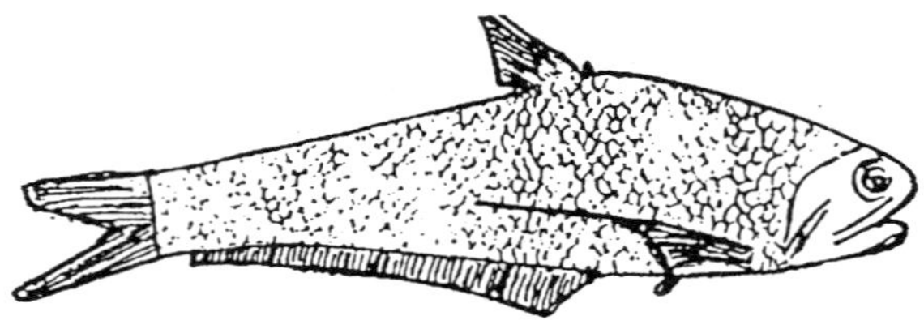

Fig. 2.35 Setipinna phasa

25. *Catla catla*

Phylum	:	Chordata
Sub phylum	:	Vertebrata
Class	:	Pisces
Sub class	:	Teleostomi
Super order	:	Ostariophysi
Order	:	Cypriniformes
Family	:	Cyprinidae
Sub Family	:	Cyprininae

Body colour is silvery with pinkish tinge. The body is short with rounded abdomen. Mouth is wide, with prominent lower jaw. Eyes are relatively large. Barbels are absent. Dorsal fin is inserted above tip of pectoral fin and gas 1-19 rays. Anal fin has 8 rays, out of which 5 are branched. Caudal fin is forked. Lateral line with 40-43 scales.

Fig. 2.36 Catla catla

26. *Rita rita*

Phylum	:	Chordata
Sub phylum	:	Vertebrata
Class	:	Pisces
Sub class	:	Teleostomi
Super order	:	Ostariophysi
Order	:	Siluriformes
Family	:	Bagridae

Body is short and compressed. Head is relatively short. Mouth is tranverse in its position. Jaws are unequal—upper jaw is longer than the lower. Eyes are small and dorso-lateral in position. Villiform teeth are present on the jaws. 3 pairs of barbels are present—one pair maxillary, one pair nasal and one pair mandibular. Dorsal fin is inserted above half of pectoral fin. Caudal fin is forked.

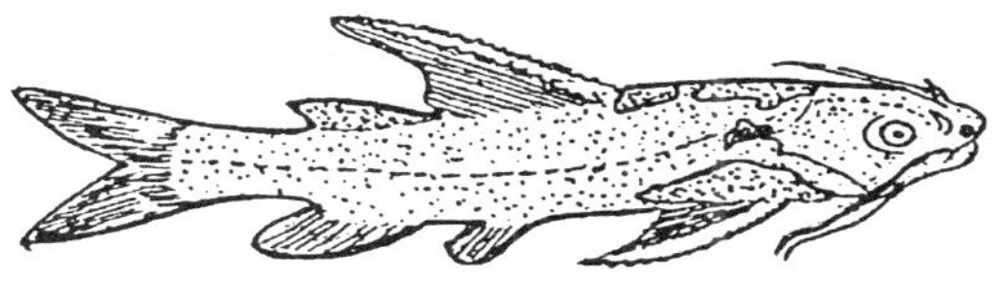

Fig. 2.37 Rita rita

27. *Macrognathus aculeatus*

Phylum	:	Chordata
Sub phylum	:	Vertebrata
Class	:	Pisces
Sub class	:	Teleostomi
Super order	:	Acanthopterugii
Order	:	Mastacembeliformes
Family	:	Mastacebelidae

Body is greenish in colour but ventral side is yellowish in colour. The body is eel-like, elongated and compressed. Head is long and pointed. Jaws are unequal, the lower jaw being shorter. Eyes are small. Small teeth present on both the jaws and also on the palate and vomer. Preorbital spines are absent. The dorsal fin is inserted far behind. Caudal fin is rounded. Pectoral fin is very short and the ventral fin is absent. There are three anal spines, second one being the longest and the strongest. Cycloid scales present. Air bladder is elongated. Lateral line is present.

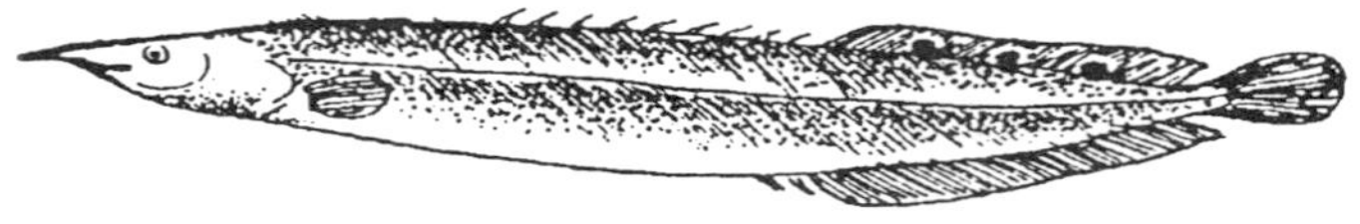

Fig. 2.38 Macrognathus aculeatus

28. *Cirrhinus mrigala* and *Cirrhinus reba*

Phylum	:	Chordata
Sub phylum	:	Vertebrata
Class	:	Pisces
Sub class	:	Teleostomi
Super order	:	Ostariophysi
Order	:	Cypriniformes
Family	:	Cyprinidae
Sub Family	:	Cyprininae

The body is elongated and compressed. The mouth is wide. The body is silvery but dark grey along back. Barbels are of very small size and are two or four. In *C. mrigala*, the dorsal fin has 15-16 rays. Lateral line has 40-45 scales. In *C. reba,* scales have dark edges. The dorsal fin is less than body height. Lateral line scales are 35-38 in number.

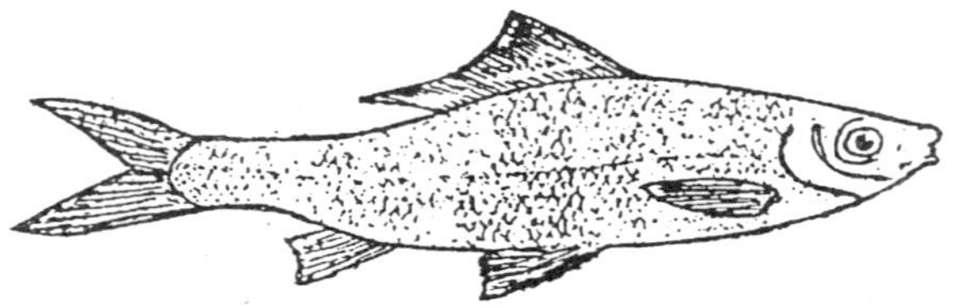

Fig. 2.39 Cirrhinus mrigala

Fig. 2.40 C. reba

29. *Eutropiichthys vacha*

Phylum	:	Chordata
Sub phylum	:	Vertebrata
Class	:	Pisces
Sub class	:	Teleostomi
Super order	:	Ostariophysi
Order	:	Siluriformes
Family	:	Schibeidae
Sub Family	:	Svhibeinae

Body is elongated. Abdomen is rounded. Head is conical in its shape. Snout is pointed. Mouth is subterminal and transverse. Jaws are unequal. Teeth are sharp. There are 4 pairs of barbels and the nasal barbels reach hind border of the head. One pair of maxillary, one pair of nasal and two pairs of mandibular barbels present. Eyes are large and lateral in position. Rayed dorsal fin is inserted above half of pectoral fin. Adipose dorsal fin is short and posteriorly free. Anal fin is long. Caudal fin is deeply forked. Pectoral and caudal fins are edged with black. Pectoral fins reach the base of the ventral. Lateral line is complete and simple.

Fig. 2.41 Eutropiichthys vacha

30. *Noemacheilus rupicola rupicola*

Phylum	:	Chordata
Sub phylum	:	Vertebrata
Class	:	Pisces
Sub class	:	Teleostomi
Super order	:	Ostariophysi
Order	:	Cypriniformes
Family	:	Cobitidae
Sub Family	:	Noemacheilinae

Body is elongated. Head is cylindrical. Mouth is inferior. Eyes are small. Jaws and palate are without teeth. Dorsal fin situated opposite the ventrals. Barbels may be 2/3/4 pairs. Anal fin is short, Lateral line is complete or incomplete.

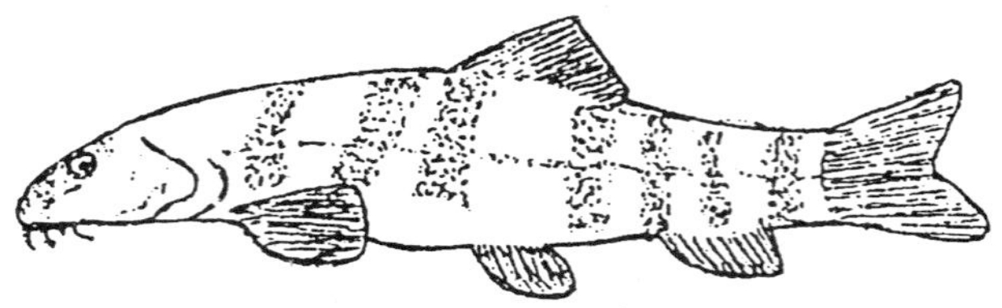

Fig. 2.42 Noemacheilus

31. *Hilsa ilisha*

Phylum	:	Chordata
Sub phylum	:	Vertebrata
Class	:	Pisces
Sub class	:	Teleostomi
Super order	:	Clupeomorpha
Order	:	Clupeoformes
Family	:	Clupeidae
Sub Family	:	Alosinae

Body is oblong and compressed. Eyes are provided with adipose lids. 15-16 pre-pelvic and 11-16 post-pelvic scutes are present in the abdomen. Head is large. Mouth is terminal. Eyes are large and lateral in position. Jaws are not equal. Dorsal fin originates opposite to the place of origin of the ventral fin. Anal fin is long and single. Caudal fin is deeply forked. The caudal lobe is as long as head. Scales are large. Lateral line is absent.

Fig. 2.43 Hilsa ilisha

32. *Amblypharyngodon mola*

Phylum	:	Chordata
Sub phylum	:	Vertebrata
Class	:	Pisces
Sub class	:	Teleostomi
Super order	:	Ostariophysi
Order	:	Cypriniformes
Family	:	Rasborinae

Moderately elongated. Mouth is anterior and the lower jaw is more prominent. Eyes are large and are centrally placed. Barbels absent. Silvery white lateral barbels present on either side of the body, which turns upwards in the tail region. The dorsal fin is inserted slightly behind the insertion of the pelvic fin and has 9 rays and no spine. Anal fin has 7 rays and the caudal fin is forked. Dark markings are present on the dorsal, anal and caudal fins. Lateral line is incomplete with 65-75 scales.

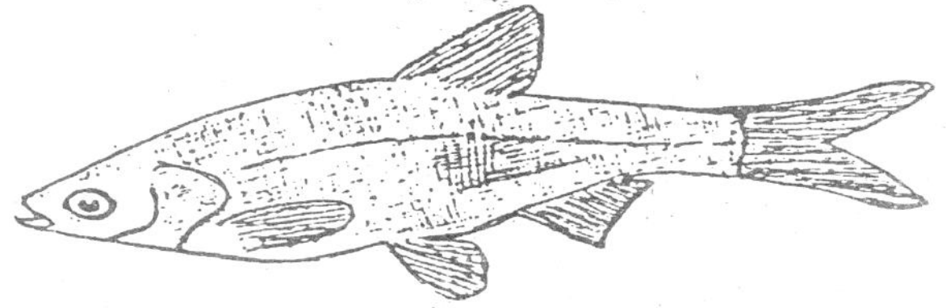

Fig. 2.44 Amblypharyngodon mola

33. *Mastacembelus armatus armatus*

Phylum	:	Chordata
Sub phylum	:	Vertebrata
Class	:	Pisces
Sub class	:	Teleostomi
Super order	:	Acanthopterygii
Order	:	Mastacembeliformes
Family	:	Mastacembelidae

Body is rich brown colour above and paler below. There is a black irregular zig zag pattern between the dorsal ridge and the lateral line. The body is fairly elongated and eel-shaped. Caudal fin is rounded. The ventral fin is absent. Air bladder is elongated. Scales are present. Jaws are unequal. Teeth are very minute and they are present on jaws and palate.

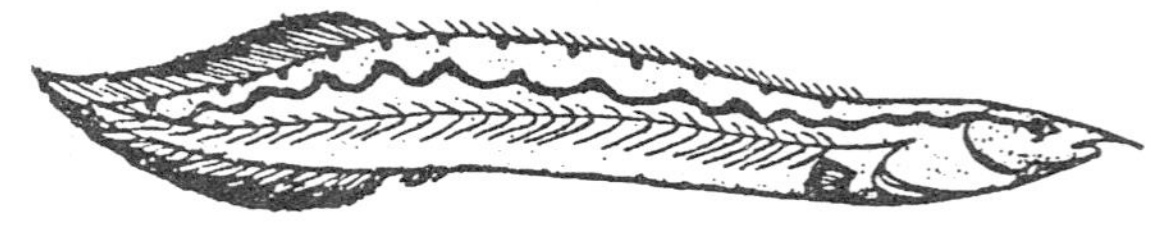

Fig. 2.45 Mastacembelus armatus armatus

34. *Crossocheilus latius latius*

Phylum	:	Chordata
Sub phylum	:	Vertebrata
Class	:	Pisces
Sub class	:	Teleostomi
Super order	:	Ostariophysi
Order	:	Cypriniformes
Family	:	Cyprinidae
Sub Family	:	Garrinae

Body is slender. Colour of the fish is olive above and dull silvery below. Dorsal and caudal fins are yellowish grey in colour, whereas the other fins are orange-coloured. Abdomen is rounded. Head is small. Post labial groove present. Pectoral fin is as long as the head. One pair of barbels are present on the rostrum. Pectoral fins are shorter than the head. Lateral line continuous with 37-40 scales.

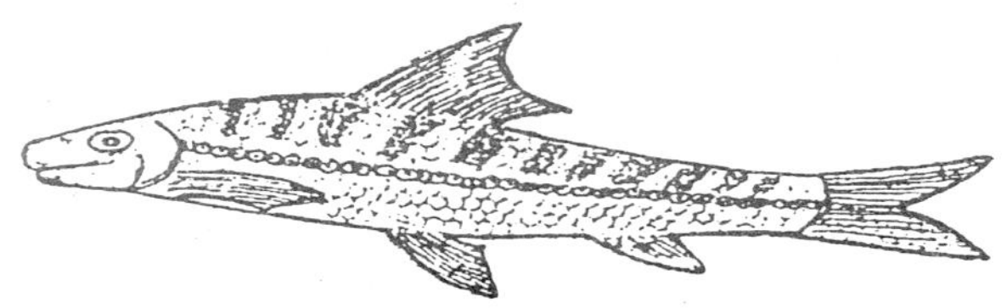

35. *Garra gotyla gotyla*

Phylum	:	Chordata
Sub phylum	:	Vertebrata
Class	:	Pisces

Sub class	:	Teleostomi
Super order	:	Ostariophys
Order	:	Cypriniformes
Family	:	Cyprinidae
Sub Family	:	Garrinae

Body is completely short. Abdomen is rounded. Head is depressed. Snout is blunt. Pores may be present on the snout. Eyes are small in the posterior half of the head. Barbels are one/two pairs. Dorsal fin is inserted ahead of pelvis. Anal fin is short. Caudal fin is slightly emarginate. Lateral fin is complete. Abdomen is yellowish green. dark spot behind the gill opening may be present.

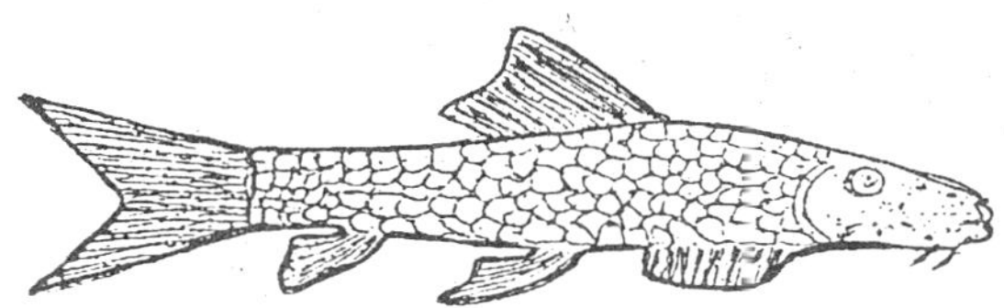

Fig. 2.47 Garra gotyla gotyla

36. *Glyptosternum maculatum*

Phylum	:	Chordata
Sub phylum	:	Vertebrata
Class	:	Pisces
Sub class	:	Teleostomi
Super order	:	Ostariophys
Order	:	Siluriformes
Family	:	Sisoridae

Body is elongated. Head is small and flat. Snout is rounded. Mouth is inferior. Eyes are small, superior in position and situated in the middle of the head. Jaws are unequal, upper jaw is larger, overhanging the lower one. Villiform teeth are present in bands on the jaws. Four pairs of barbels are present—one pair of nasal, one pair of maxillary and two pairs of mandibular. Dorsal fin is inserted above tip of pectoral fin and has 6-7 rays and there is no spine. Pectoral fin has 11 branched rays and there is no spine. Pelvic fin has only 6 rays. Caudal fin is rounded. Anal fin is short and has 5 or 6 rays. Lateral line is simple and complete.

Fig. 2.48 Glyptosternum maculatum

Fig. 2.6 G. reticulatum

37. *Puntius chola*

Puntius sarana sarana and

Puntius sophore

Phylum	:	Chordata
Sub phylum	:	Vertebrata
Class	:	Pisces
Sub class	:	Teleostomi
Super order	:	Ostariophysi
Order	:	Cypriniformes
Family	:	Cyprinidae
Sub Family	:	Cyprininae

Body is moderately elongated and deep. Head is short. Snout is obtuse and conical. Mouth is arched. Jaws are simple. Eyes are moderate to large in its size and dorso-lateral in position. Barbels are one pair. Dorsal fin is short and is inserted nearly oposite pelvic fins. Anal fin is short. Caudal fin is forked. Scales are generally small. Lateral line is complete.

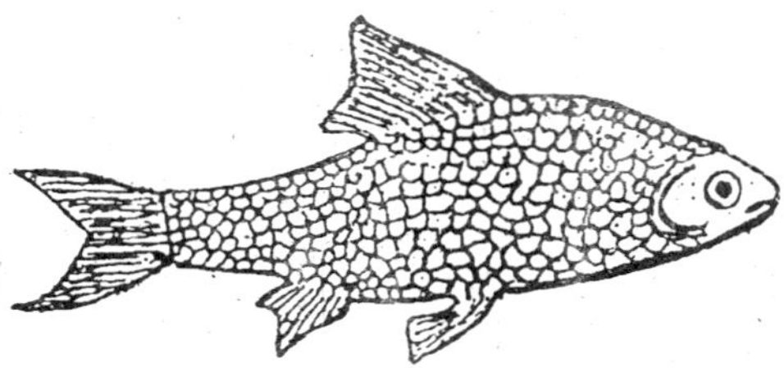

Fig. 2.50 Puntius chola

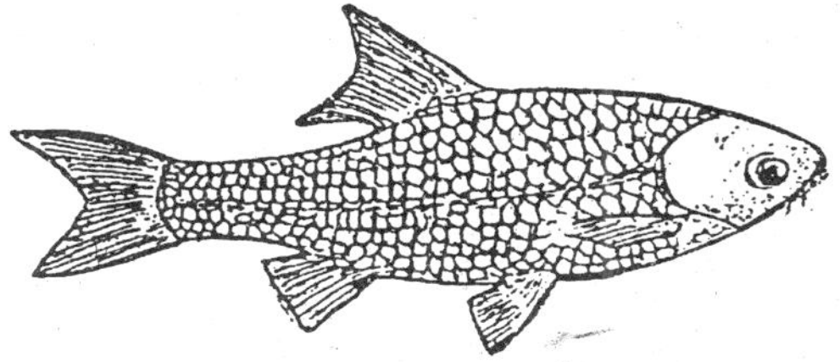

Fig. 2.51 P. sarana sarana

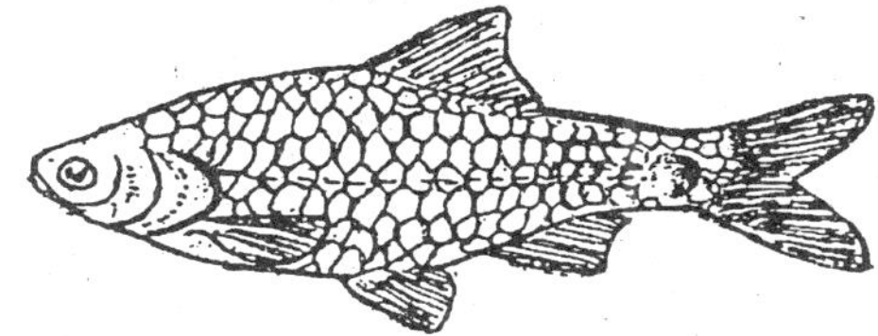

Fig. 2.52 P. sophore

38. *Colisa faciata*

Phylum	:	Chordata
Sub phylum	:	Vertebrata
Class	:	Pisces
Sub class	:	Teleostomi
Super order	:	Acanthopterygii
Order	:	Perciformes
Family	:	Belontidae
Sub Family	:	Trichogastrinae

Body is elevated and laterally compressed. Snout is blunt. Jaws are unequal and are slightly protractile. Villiform teeth present over the jaws. Branched arches have toothed tubercles. A dorsal fin commences above from near pectoral base. Caudal fin is truncated. Dorsal, caudal and anal fins are spotted orange in colour. Pelvic fin will be in the form of elongate, filiform ray. Ctenoid scales are present. Eyes are large.

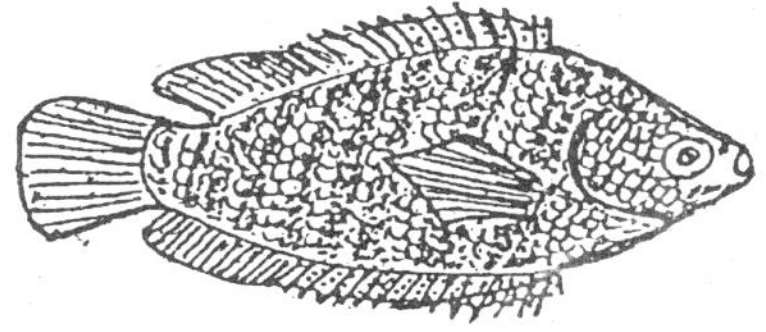

Fig. 2.53 Colisa faciata

39. *Gadusia chapra*

Phylum	:	Chordata
Sub phylum	:	Vertebrata
Class	:	Pisces
Sub class	:	Teleostomi
Super order	:	Clupeomorpha
Order	:	Clupeoformes
Family	:	Clupeidae
Sub Family	:	Alosinae

Body is oblong, compressed and covered with small scales. Abdomen is serrated with 18-19 pre-pelvic and 8-10 post pelvic scutes. Barbels are absent. Head is short. Mouth is terminally placed. Eyes are large and lateral in position. Dorsal fin is inserted above pelvic origin with 14-15 rays. Caudal fin is forked and the anal fin has 21-24 rays. Lateral line is absent. Teeth are absent.

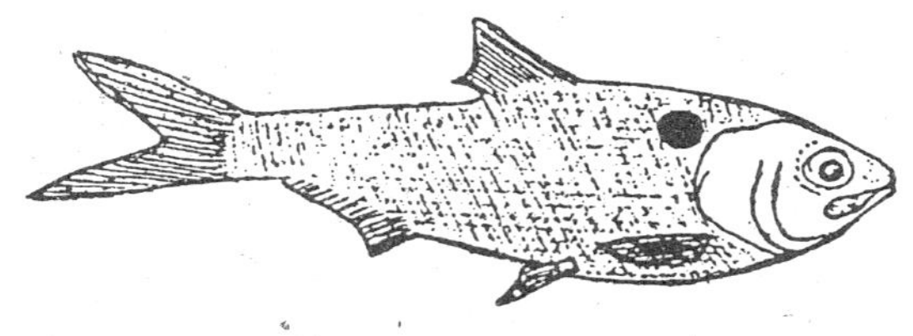

Fig. 2.54 Gadusia chapra

40. *Bagarius bagarius*

Phylum	:	Chordata
Sub phylum	:	Vertebrata
Class	:	Pisces
Sub class	:	Teleostomi
Super order	:	Ostariophysi
Order	:	Siluriformes
Family	:	Sisoridae

Body is elongated. Head is large. Snout is conical. Mouth is ventral and wide., Eyes are small and dorsally placed. Barbels are 4 pairs. The maxillary barbels have broad base and is longer than the head, reaching anterior one third of pectoral fin. Jaws are unequal. Teeth are present over jaws. The dorsal fin is rayed and inserted above the base of dorsal fin. Adipose dorsal fin is moderately long, and posteriorly free. Caudal fin is deeply forked. The upper lobe of the caudal fin is longer. Lateral line is simple and complete.

Fig. 2.55 Bagarius bagarius

41. *Mugil cephalus*

Phylum	:	Chordata
Sub phylum	:	Vertebrata
Class	:	Pisces
Sub class	:	Teleostomi
Super order	:	Actinopterygii
Order	:	Mugiliformes
Sub Order	:	Mugiloidae
Family	:	Mugilidae

Body is normally compressed. Head is broader. A band of teeth present on both the jaws. Body is greyish brown on back, silvery on sides and whitish ventrally. They feed generally on diatoms, blue-green algae, green algae, large aquatic plants and crustaceans and also zooplankton. These fish mature at a size ranges from 250-400 mm. Males mature in first year and females in second year. Breeds in off-shore. Eggs need seawaters of high salinities for fertilization.

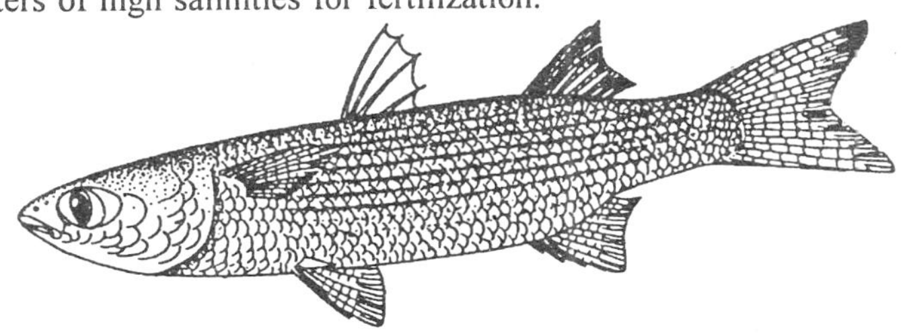

Fig. 2.56 Mugil cephalus

42. *Chanos chanos*

Phylum	:	Chordata
Sub phylum	:	Vertebrata
Class	:	Pisces
Sub class	:	Teleostomi
Super order	:	Clupeomorpha
Order	:	Clupeoformes
Sub Order	:	Chanoidae
Family	:	Chanidae

Chanos body is elongated and compressed. Mouth is relatively small, anterior and transverse in position. Upper jaw overhanging lower jaw. Pectoral fin pointed with elongated scaly appendage at base. Capable of tolerating wide variation in salinity (0-40 ppt).

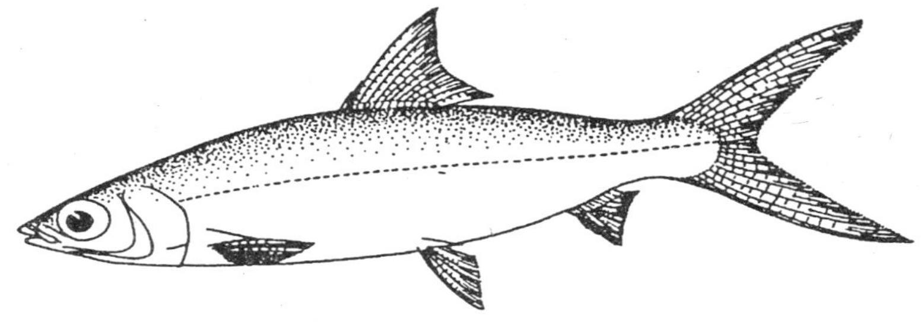

Fig. 2.57 Chanos chanos

43. *Elops machnata*

Phylum	:	Chordata
Sub phylum	:	Vertebrata
Class	:	Pisces
Sub class	:	Teleostomi
Super order	:	Clupeomorpha
Order	:	Clupeoformes
Sub Order	:	Clupeoidei
Family	:	Elopidae

The body of this fish is elongated. Head is conical and snout longer than eye. Maxilla extending beyond eye and lower jaw projecting. Body colour is silvery and fins yellowish with greyish tinge. This fish generally feed on crustaceans.

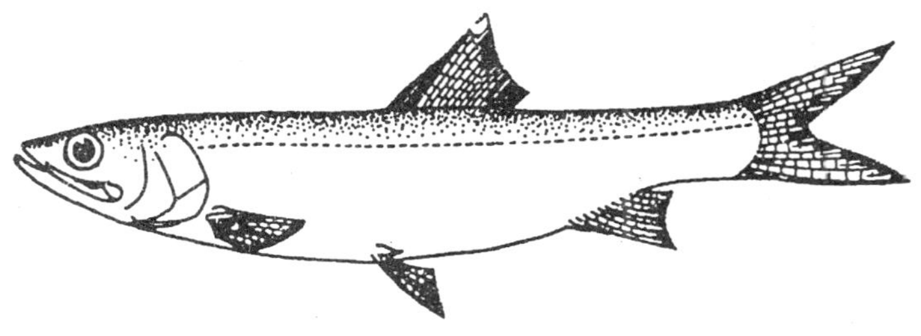

Fig. 2.58 Elops machnata

44. *Megalops cyprinoides*

Phylum	:	Chordata
Sub phylum	:	Vertebrata
Class	:	Pisces
Sub class	:	Teleostomi
Super order	:	Clupeomorpha
Order	:	Clupeoformes
Sub Order	:	Clupeoidei
Family	:	Megalopidae

The body is compressed. Maxillary reaches to opposite hind edge of eye. Villiform teeth present over jaws, vomer, palatines, pterygoids and sphenoids.

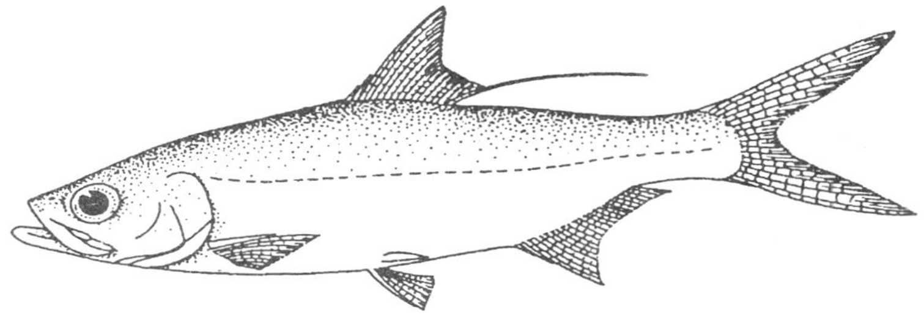

Fig. 2.59 Megalops cyprinoides

45. *Etroplus suratensis*

Phylum	:	Chordata
Sub phylum	:	Vertebrata
Class	:	Pisces
Sub class	:	Teleostomi
Order	:	Perciformes
Sub Order	:	Percoidae
Family	:	Cichlidae

Commonly known as pearl spot. Body is oblong and compressed. Lower jaw slightly longer than upper jaw. Feeds chiefly on filamentous algae and also zooplankton, shrimps, insect larvae, gastropods, insects, bivalves, mysids, sponges. Tolerate wide variations of salinity from 0-40 ppt.

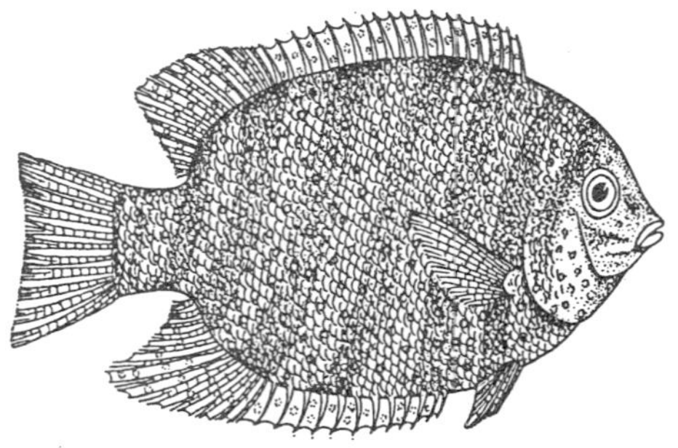

Fig. 2.60 Etroplus suratensis

46. *Lates calcarifer*

Phylum	:	Chordata
Sub phylum	:	Vertebrata
Class	:	Pisces
Sub class	:	Teleostomi
Order	:	Perciformes
Sub Order	:	Percoidae
Family	:	Centropomidae

Body is moderately compressed. Mouth is slightly oblique. Maxilla extends to below the hind edge of orbit. Lower jaw projecting. Villiform teeth present over jaws, vomer and palate. The fish is capable of tolerating wide variations in salinity (0-40 ppt).

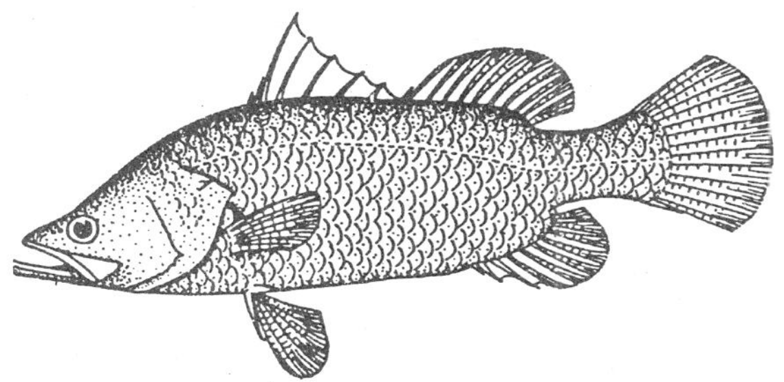

Fig. 2.61 Lates calcarifer

47. *Eleutheronema tetradactylum*

Phylum	:	Chordata
Sub phylum	:	Vertebrata
Class	:	Pisces
Sub class	:	Teleostomi
Order	:	Polynemiformes
Family	:	Polynemidae

Commonly known as Indian Salmon. This fish has large mouth reaching far behind eye. Villiform teeth present. Feeds on prawns, amphipods, isopods, stomatopods, etc. Male matures when reaches 225 mm and female 285 mm.

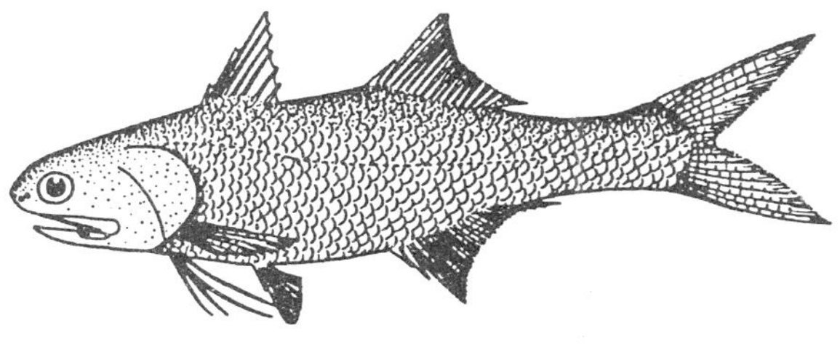

Fig. 2.62 Eleutheronema tetradactylum

3

GENERAL BIOLOGY OF FISH AND PRAWN

The body shape and function of a fish is totally adopted to a free-swimming life in water, the body shape being adapted to give maximum efficiency to its propulsion through the water. Its internal organs, whilst performing the same basic functions as those of a man, are adapted to perform those functions in water. There are basically two shapes of fish, round fish and flat fish. Flat fish, such as skate, ray, plaice and sole are adapted to life on the bottom of the water. Round fish, in general have evolved an efficient hydrodynamic shape to allow them to move through the water body with the minimum expenditure of energy. The animal's internal skeleton is used to form the frame to which are attached the muscles used for swimming and breathing. The fish propels itself through the water by sinusoidal movements of the body amplified by the flat tail. The fins attached to the side of the body are used for stability and to change direction in the same way as the wings of a sliding bird. The outer surface of the body is covered by such. There are basically two types of fish skin, those covered in little bony plates called scales, and those without scales. Both types of skin secrete a mucus which makes the fish slippery to handle. This mucus has three main functions, it helps the passage of fish through the water by reducing the friction of the skin, it helps to water proof the fish, it gives protection against attacks of bacteria and other microscopic organisms which otherwise infect the fish. It is therefore very important that fish are handled with care to avoid damage to the skin.

A fish breathes by extracting oxygen from the water in which it is living by means of gills. The gills are vascular organs situated behind the head and protected by a bony cover called operculum. The gills function in much the same way as do our lungs, oxygen is extracted from the water, or from the air in the case of lungs, across a wet surface and into the blood stream of the animal. There is, however, one very important difference between air-breathing and water-breathing. Fish are faced with two different situations depending on the salinity of the water in which they are living. Freshwater fish live in an environment which is much more dilute than is their body fluid. Therefore, as salt attracts water from the air so the fishes body fluid attracts water and the fish live in constant danger of 'drowning' by an excessive intake of water into the body tissues. Seawater fish on the otherhand live in water which is much more concentrated than is their body fluid and they face the danger of dehydration as water tends to leave their bodies. Fish catch their food in their mouths, which are equipped with teeth. The food is not, however, chewed, although some plant-eating species have grinding plates in the mouth cavity which are used to macerate the vegetable matter eaten by the fish. The food is swallowed and digested in an alimentary canal. Fish have well developed sensory organs. Fish have eyes, but no eyelids. They have internal ears and balance organs but no external ear. No nose, but are sensitive to the chemical content of the water and can 'smell' the chemicals. The organs of 'smell' are distributed over the skin but are also concentrated in pits in the head which are very similar to our nostrils. A fish has a well-developed nervous system with a brain and spiral column, spinal nerves and cranial nerves.

A female fish lays eggs, which are fertilized by sperm produced by the male. The eggs and sperm in the great majority of species are released onto the water in which the fish are living. Fertilized eggs either float in the water body or fall to the bottom. Bottom eggs are sometimes guarded in nests by the fish or else they are sticky and become attached to weeds and stones. The eggs hatch in a few days or weeks depending on the species and the water temperature. The young fish hatch with small amounts of yolk attached to them which supplies them with the food they need during their first days of life. Some fish guard their young, some even guard them in their mouths. Other little fish have to fend for themselves.

The various problems encountered by a fish living in the sea, such as water temperature, levels of dissolved oxygen, and the high salt content of sea water, are all problems for crustaceans also. The most immediately obvious thing about a crustacean is the fact that it is covered on the outside by a hard surface, which is hardest in the crabs and lobsters. This hard outer shell is in fact continuous over the whole of the animal and is formed by a series of hard rings which are joined together by this flexible outer layer. This hard outer layer creates one obvious problem for the animal, namely, how to grow in size. The crustacean solves this problem by periodically casting the whole of its outer hard layer. The shell is cast and at this stage the animal is covered by a soft flexible skin. During this period the animal absorbs water and grows rapidly. After it has swelled the shell is hardened by the deposition of calcium salts so that in a few hours the animal has regained its former state but is larger. During the young stages of the animal these moults occur very frequently and often at the moult the animal changes its form. This change is called a metamorphosis. The Decapoda, lobsters, crabs, and shrimps have a simple alimentary canal consisting of a straight tube with two anterior chambers. They have a simple blood system, a rudimentary heart and simple nervous and excretory systems. The sexes are separate, the male fertilizing the female, which stores the sperm and releases them on laying her eggs. Crustaceans have a complicated life cycle during which they pass from the egg stage to the adult through many different shapes and larval forms.

FISH

Skeletal and muscular system

Skeletons of fish are of two types, cartilaginous and bony. Cartilaginous skeletons never become calcified and therefore remain soft. These are the types of skeletons which skates, rays and sharks have. All other fish lay down calcium salts in their skeletons to form bone of varying degrees of hardness.

The function of the skeleton is to form a physical support for the body and a base against which the swimming muscles can pull. As the fish is living in water the body is supported by the water and so the function of skeleton, unlike that of a land animal, does not include the necessity to support the body of the animal against the force of gravity.

The skeleton consists of a skull case which protects the brain, fastened to which is a flexible spinal column built up of vertebrae. From the backbone there are bones, a little like our ribs, against which the powerful swimming muscle can pull. Most of the flesh of the fish consists of these muscles which are located on either side of the vertebral column. The muscles are alternately flexed and contracted so as to cause the fish's tail to move from side to side and thus to push it forward. There are bones which support the fins and the tail and various small bones in the muscles which aid the function of the muscles. Attached to the skull are jaws . Behind the skull is a bony plate called the operculum which protects the gills and also, when moved open and closed, helps pump water over the gills.

The skin of most fish is covered, to a greater or lesser extent, with small bony plates called scales. These plates are not, as is commonly supposed, outside the skin but are embedded in the skin and so cause damage if they are rubbed off. The scales grow in size as the fish grows. The scales tend to grow at different rates at different times of the year varying with the changing growth pattern of the fish. Changes in the growth pattern of the scales are marked by rings in the scale structure. Because of these marks the scales can be used to determine the age of the fish.

Circulatory system

In the section on fish nutrition it is explained that in order to release the energy contained in its food a fish requires oxygen. The oxygen for this process is extracted from the water by the gills and is transported from the gills to the tissues by the blood. In the blood the oxygen is carried in combination with a red chemical containing iron called haemoglobin which is carried through the blood in red blood cells. The blood is directed round the body in the blood vessels through which it is pumped by the heart. In man the heart has four chambers; the blood passes from the lungs to the heart and is then pumped from the heart to the body. In a fish the heart is much more simple and has only two chambers, which are 'in line' in the main blood vessel and thus act as a straight pump. The blood returns to the heart from the body, is pumped to the gills for oxygenation and then continues on to the body. The gills are situated at the side of the head behind the eyes and are covered by the opercular plate. The gills are feathery structures, very thin-skinned and containing many blood vessels. As the water is passed over the gills carbon dioxide, a product of the metabolism of the fish, diffuses from the blood into the water and oxygen diffuses from the water in to the blood. Water is actively pumped over the gills by the fish using the muscles of the mouth and the operculum.

Digestive system

The gut is a tube, which stretches from the mouth to the anus. It is divided into four parts:

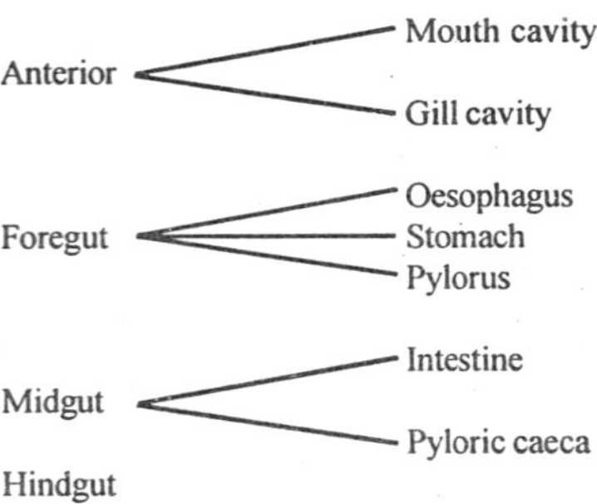

The mouth cavity and jaws are used for capturing or taking up food. There are teeth around the jaw edge and backward-facing teeth inside the mouth; these are very well developed in fish which eat other fish to enable them to hold the prey. Inside the mouth and on the bars which support the gills are grinding plates; these are well developed in fish which eat vegetable matter. The stomach varies considerably in size and shape depending on the type of food eaten. For example, continuous mud-eaters have hardly any stomach, whilst fish-eaters have large stomachs in which to retain their prey for initial digestion. The passage of the food from the stomach to the pylorus is controlled by a muscular ring called the pyloric sphincter.

The digestive processes of the midgut have not been extensively studied; it is thought that the pyloric caeca have no special function but apparently merely serve to enlarge the surface area of the gut.

The hindgut continues the process of water and food absorption. The total length of the gut differs according to species, depending on the type of food they eat. Thus detritus eaters have the longest gut and carnivores the shortest.

The processes of digestion have been little studied and the times of digestion are not well known. As a fish is cold-blooded, times of digestion are much longer than in a warm-blooded animal and it may take days for a fish to digest a meal. There does not appear to be any digestive enzyme produced in the mouth of a fish and digestion appears to start in the stomach. Digestive enzymes are chemicals produced by the body which are passed in to the lumen of the gut, where they breakdown food into its component parts, which can then be absorbed through the gut wall into the blood stream. In a fish the stomach produces an enzyme called pepsin which digests protein. The midgut produces trypsin which further digests protein, lipase which digests fats and amylases which digest carbohydrates.

The liver appears to function in the same way as in other vertebrate animals. It is the chemical factory of the body. Here waste products of the body are processed for excretion and the products of digestion are processed for use. As in other vertebrates the liver discharges a liquid in to the gut called bile which contains some waste products from the body. It also emulsifies fats so that they can be digested more easily. The active components of bile are bilirubin and biliverdin which are the products of the breakdown of haemoglobin, the red pigment of the blood.

Excretory system

The products of the functioning of the fish's body and the products of the 'burning' of food to produce energy are got rid of by the body in two ways, via the excretory system. The kidney's, which are the main organs of the excretory system, are large and are situated along the top of the abdominal cavity. The kidney has two important functions. It excretes waste products from the body and it plays an important role in maintaining the equilibrium of the body fluids. Freshwater fish take up water from their environment. Whilst sea fish lose water to their environment. In order to maintain the composition of their body fluids freshwater fish excrete a copious dilute urine, the kidneys retaining the dissolved salts within the body. Fish living in the sea have to drink water in order to survive since they are in constant danger of dehydration as they lose body water. Drinking water allows the fish to maintain its water content but it means that it also takes in unwanted levels of sea salts. A marine fish gets rid of unwanted chloride via its gills and its kidney excretes a urine equal in salt content to the sea water.

Some fish, such as salmon, migrate from sea to freshwater and have kidneys which can adapt to either condition. The kidney of a fish is not as well developed as is that of a land vertebrate. One result of this more primitive kidney is that a fish cannot excrete various nitrogenous compounds which a land vertebrate can. These compounds are deposited under the skin in the form of a chemical substance called guanine.

Guanine reflects the light and appears silvery, hence the silver colour of the fish.

Endocrine organs

In a vertebrate animal the awareness of the outside world is transmitted to the brain via the sense organs of touch, smell, taste, sight, temperature and vibrations. These sensations are transmitted to the brain by the nerves and are immediately reacted to in various ways. The animal also reacts to its environment in a slower way, this reaction being brought about by the secretion of chemicals in to the body by endocrine organs. These chemicals are called hormones and effect the long-term reaction and development of the animal, sexual maturity and growth being examples. A fish has nearly the same endocrine organs as other vertebrates. Obviously their function is of great importance to the fish farmer.

This small gland, located underneath the brain has been described as the 'master gland' of the body. The hormones produced by the pituitary affect the fish directly but also it produces hormones which act via the other endocrine organs. Thus the pituitary, which receives its signals direct from the brain, controls the reaction of the other endocrine organs to the environment and hence to a greater or lesser extent the actual activity of those organs. The pituitary produces the growth hormone which is believed to have a direct effect on the growth of the body as it has in many vertebrates. Nevertheless, its exact function is not fully understood. Obviously it is important for farmers to be aware of the growth hormone.

The thyroid is situated under the gills along the main blood vessel which runs between the heart and the gills. In man the thyroid controls the rate of metabolism, the rate at which the engine runs, in fact. The function of the thyroid in fish is not yet fully clear although there is some evidence to suggest that it also has an effect on the metabolism.

Gonads

Fish are either male or female and as such have either testes or ovaries. The gonads lie in the body cavity. Those of many marine fish are eaten as roe and the sturgeon's form the much prized caviar. As well as producing eggs or sperm the gonads also secrete hormones into the blood stream. These hormones are responsible for the differences in appearance and behaviour between the males and females. The ripening of the gonads and the spawning activities of the fish which result from the hormones secreted into the blood stream by the gonads are, in the main, the result of changes in the fishes external environment. These changes act via the sense organs to the brain, from the brain to the pituitary and from the pituitary via hormones to the gonads. In this way changes outside a fish cause changes inside it which in turn can cause changes in activity. A fish egg is composed of a tough outer shell, the embryo and a yolk store for the nutrition of the developing fish. The size of the egg and the amount of yolk vary with the species. In most tropical species the female lays many small eggs, perhaps 1 mm in diameter, which quickly develop and hatch within a few days of being laid. Salmon and trout, on the other hand, lay fewer but larger eggs, about 5 mm in diameter with larger yolk stores. These eggs take longer to hatch and the larvae have large yolk sacs attached to them on hatching.

Application of hormones in farming

In recent years two very important application of hormones have been developed in fish farming. The first concerns the use of pituitary hormones to induce a fish to spawn. It has been found in practice that many fish, such as the mullet and grass carp, which are of great importance in farming, will not spawn naturally under farm conditions. Grass carp, for example, need to complete long migrations before their gonads ripen. Obviously this is an effect of the external environment on the physiological stage of the fish. However, it has been discovered that if adult fish nearing maturation are given injections of pituitaries taken from other ripe fish the recipient fish will become ripe and will spawn under farm conditions. This procedure has been further refined by the use of injections of hormones and not whole pituitary extracts. The second use of hormones concerns the control of population growth in *Tilapia* which by excessive breeding can result in farm ponds being super populated with small fish of low market value. Obviously one way of controlling this would be to rear unisex populations. Various ways have been tried to obtain such populations and recently a simple effective system has been developed. This technique depends entirely on the use of the hormones produced by the gonads to control the sex of the fish. It has been found that if very young fish are dosed with male sex hormones an all male population results, and this technique is now being successfully utilized with various species.

Nervous and sensory systems

Fish have a well-developed nervous system with a complicated well-developed brain and spinal cord. There are nerves from the spinal cord to the body as well as nerves directly from the brain which perform special functions. The nervous system is in fact comparable to that of other vertebrates. The sense organs of an animal are those parts of the body which allow the animal to be aware of its external environment and changes in that environment. These organs can be classified according to their function as follows: photoreceptors—sensitive to light; chemo and thermoreceptors—sensitive to changes in water chemistry and temperatures; mechanoreceptors—sensitive to touch and vibration.

Fish have a pair of eyes as photoreceptors which are constructed in the general plan of vertebrate eyes. That is to say they have a transparent frontal cornea, and an internal lens which focuses the received image onto a light-sensitive retina at the back of the eye. The eye can be turned by eye muscles to allow the fish to follow images. The focusing mechanism of the fishes' eye is different to that of man in that little change in the shape of the lens occurs but, in order to the focus, the lens is moved in respect of the retina. A fish has three types of chemoreceptors, an olfactory organ with which it smells, taste buds, and a general set of sensory cells which are scattered over the skin all over the body with which it detects changes in the water chemistry. The olfactory organs are situated in closed pits on the upper side of the head in front of the eyes. In sharks which have a keen sense of smell for hunting their prey, the pit is open to the mouth and the water current passes through it. In other fish the pits are close to the mouth but generally have two openings to the exterior so that the water passes through the pit. The taste organs are situated in the mouth and also, except in sharks, similar organs are situated all over the fish's body. The general sensitivity of the fish to changes in the chemistry of the water is also due to a scattering of nerve endings which occur all over the body surface. A fish is therefore very sensitive to the chemistry of the water in which it is living and can detect very small changes in that chemistry, salmon fry, for example, can detect some chemicals at a dilution of 1×10^{11} This sensitivity to dissolved substances is of importance to the fish farmer who must always remember the awareness of fish to chemical changes in the water.

Fish are very sensitive to changes in water temperature, again important to remember as farmers may needlessly subject fish to sudden changes in water temperature. The temperature receptors are situated in the skin and are scattered over the body surface. It has been demonstrated that fish can be sensitive to changes in temperature as low as 0.03°C.

Mechanoreceptors include organs sensitive to sound waves and vibrations, including water movements. A fish has two types of mechanoreceptors, the inner ear and the lateral line. A fish possesses an inner ear on each side of its head similar in structure to the inner ear of man. These structures are sensitive to vibrations of the frequency of sound waves travelling through the water. The second system is called the lateral line system. This runs down the length of the body, half way up the side. The lateral line is a canal-like structure situated in the skin, containing sensory cells equipped with hairs which project out into the canal. The hairs bend in water currents and hence are sensitive to vibrations and to water movements. A fish farmer should be aware that his animals are very sensitive to sound and to vibrations. Indeed, as sound travels five times faster in water than it does in air, the fish in a pond may hear a sound before the farmer. The inner ear, as in other vertebrates, consists of semicircular canals connected to a central chamber. In the central chamber there is a small flat bone called an otolith. As is explained later, under aging of fish, this otolith is used to determine age. The inner ear lining contains sensory hair cells which detect sound vibrations, head movements and also the position of the fish relative to the pull of gravity. This organ of balance is therefore important to a fish for it is often the case that a fish in deep or muddy water has no reference point when it cannot see the bottom. Without the inner ear it would not know which way up it was.

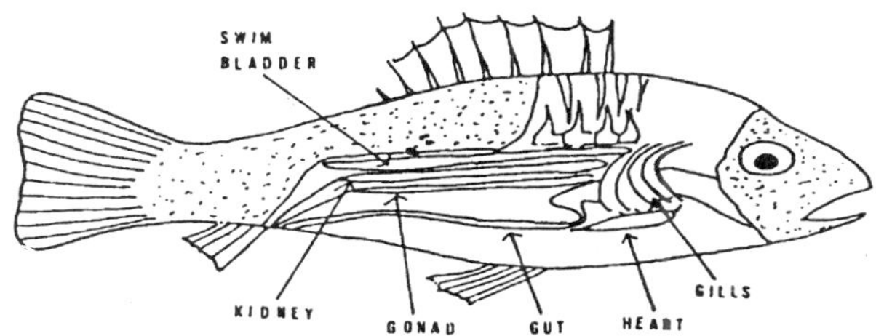

Fig. 3.1 Schematic general anatomy of a fish

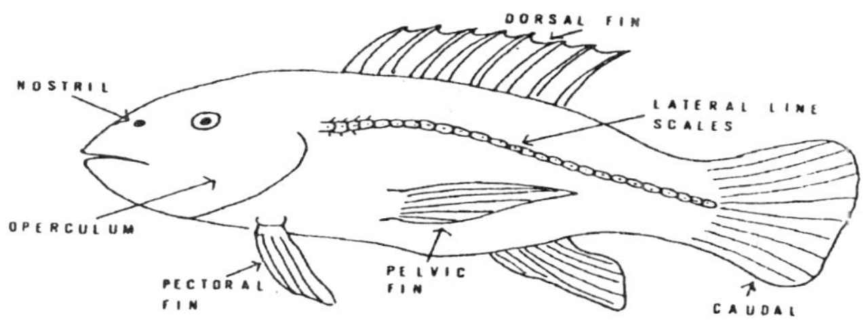

Fig. 3.2 General external features of a fish

CRUSTACEANS

Skeletal system

Crustaceans have no internal skeleton like that of a fish but they have evolved an external one, called an exoskeleton, which, when heavily armoured, as in crabs and lobsters, gives the animal a degree of protection from its predators. In the Malacostraca, the sub class to which belongs the order Decapoda, which includes all the animals farmed, the animals have bodies divided into 19 segments. The exoskeleton is also divided into 19 segments. Between the segments there is no deposition of calcium salts so that at this point the cuticle forms a flexible hinge. Each segment bears a pair of appendages which are specialized to perform a particular function. At the sides of each segment there is an interlocking joint which means that, unlike a fish, a shrimp can swim only by flexing the tail up and down and not from side to side. The body can be divided into three regions, the head, the thorax and the abdomen. The head is composed of five segments, the first two bearing the antennae and the last three the appendages used for eating. The next eight segments form the thorax, the last five of which carry pairs of walking legs, from which is derived the name of the order Decapoda, meaning ten feet. In the decapods which concern us, the head and thorax are joined to form a cephalothorax, covered by a hard carapace. The last section of the body, the abdomen, is composed of six segments and each segment carries a pair of appendages used for swimming. To be able to grow, a crustacean has first to cast its hard outer skeleton. It grows a new soft skeleton inside its hard old skeleton which then splits along the epimeral line and the animal withdraws. It then absorbs water and swells in size, the new soft outer cuticle then forms a new hard exoskeleton and the animal has grown. There is no internal skeleton as such but the exoskeleton bears inwardly pointing plates and ridges to which the muscles are attached.

Muscles

The muscles of a shrimp and a lobster can be divided by their function into two types. The main body of the muscles run the length of the tail and are used to flex the tail in swimming. These muscles are the only ones of commercial importance in these animals. In the lobster and crab the muscles inside the walking legs and the powerful princers may also be extracted and eaten.

Circulatory system

In the Decapoda, the heart is a simple tube which lies along the dorsal (upper) surface of the thorax. From the heart the blood is circulated to the body by a system of arteries. There is no vein system returning the blood from the body to the heart. Instead the blood flows through a series of spaces in the body to a space surrounding the heart, called the pericardium. Blood enters the heart from the pericardium through six valves in the heart wall which allow the entrance but not the exit of the blood. As it returns to the heart, some, but not all, of the blood passes through the gills where gaseous exchange between the blood and the water occurs. This system is not as efficient as that of a fish as not all of the blood passes through the gills on each cycle. The gills are situated under the hard carapace. The gills are thus protected and are situated in a confined space through which the animal can pump water to increase the efficiency of respiration. The respiratory pigment of the Decapoda contains copper. It is blue in colour and is called haemocyanin. The haemocyanin is not carried in the blood in special cells but is dissolved in the blood fluid.

Digestive system

The gut can be divided into three parts, the foregut, the midgut and the hind gut. The foregut consists of two chambers with folded walls which carry calcified plates. The animal tears its food with the feeding appendages of the head and passes it through the mouth into the foregut, where it is ground up by the calcified plates. The midgut is the region which contains various tubules opening from the gut wall and also a large glandular structure which is called the hepatopancreas, the function of which, although, it is known that it is connected with the digestive process, is not fully understood. The hindgut stretches from the midgut down the length of the tail to the anus, its function is also not well understood.

Excretory system

The crustaceans do not possess organs comparable to the vertebrate kidney, excretion appears to be carried out by a variety of glandular tissues, the most important of which are situated in the head and associated with the antennae and the eating appendages.

Reproductive system

The sexes are seprate. The gonads are situated in the thorax and consist of a pair of hollow organs which communicate with the exterior by ducts. The male forms the sperm into discrete lumps called spermatophores, which are attached to the female and release the sperm when she lays her eggs.

Endocrine system

The crustacea have a well developed endocrine system. Three sets of glands have been identified in the Decapoda, the X-organ-sinus gland system, the pericardial organ system and the post commissural organs.

X-organ sinus gland system

These glands are situated in the eye stalks and are endocrine organs of great importance to the aquaculturist. This is the major endocrine organ complex and is in many ways probably equivalent to the vertebrate pituitary. At present four hormones have been identified as being produced by this gland which it is suggested, have about 14 different functions in the physiology of the animal. Of these, two are of importance to the farmer. These are the hormone which controls moulting and that which controls the ripening of the gonads. The existence of this latter hormone is the reason why removal of an eye stalk has been found to affect the sexual maturation of the animals, and so is used as the basis of the technology to obtain ripe animals under farm conditions.

Pericardial organs

These lie in the pericardial cavity and produce hormones, but their function is unknown.

Post-commissural organs

These lie in close proximity to the nervous system. Their function is not understood although it is known that they produce one hormone concerned with the pigment of the animal.

Nervous and sensory systems

The decapods have a well-developed nervous system consisting of a brain-like concentration of nervous tissue in the head and a ventral nerve cord which runs the length of the body. The brain is composed of three parts and sends nerves to the feeling appendages, the eyes and the antennae. Pairs of nerves leave the nerve cord in each segment and innervate the various parts of the body.

The decapods have a pair of eyes as photoreceptors, situated at the end of the eye stalks on the head. These are unlike vertebrate eyes but are similar to the compound eyes of insects. The vertebrate eye has one lens and one retina on to which the received image is focused by the lens. The whole eye can be moved to follow moving objects. The compound eye is composed of many very small eye-like structures. The whole eye, which cannot be moved, is covered with a transparent cornea, a continuation of the body cuticle under the cornea are many cone-shaped organs which form the facets of a sphere-shaped eye. Each separate organ is lined with a retina and connected to the brain by a nerve the definition, the clarity of vision, of a compound eye such as this is very difficult to determine but it is very good at perceiving movement as the image passes from one facet of the eye to the other, a fact which is obvious when one tries to catch a fly.

Aquatic crustaceans posse well-developed chemo and thermo receptors. These organs are scattered at different points in the external cuticle of the animals, most of them being in the form of hair like protrusions. These receptors are very sensitive and respond to chemical substances, temperature and pH (alkalinity and acidity). They are also used to locate members of the opposite sex. Some crabs can detect food extracts in the water at dilutions of 10^{11}. The sensitivity to temperature changes is not very great and the animals can detect changes of the order of 1°. The animals' sensitivity to changes in the chemistry of the water is important to shrimp farmers as this will affect their ability to detect food and also their reaction to unfavourable conditions.

The crustaceans have mechanoreceptors present in the cuticle which respond to touch and to water movement. In the Decapoda there is an organ of balance at the base of the small antennae. This organ consists of a pit lined with sensory hairs and containing sand grains. The position of the sand grains on the sensory hairs allows the animal to be aware of its position relative to the pull of gravity. This fact can be demonstrated by giving a newly moulted animal with iron filings instead of sand; when these enter the pit the pull of a magnet causing the iron filings to rise up will cause the animal to swim upside down.

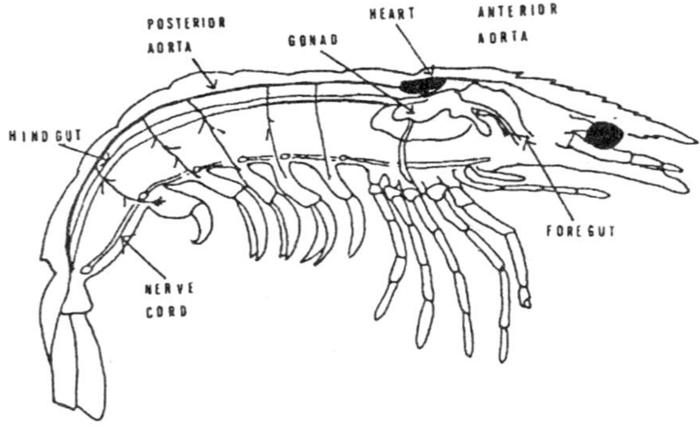

Fig. 3.3 Main internal organs of a penaeid shrimp

4

FOOD AND FEEDING

FOOD AND FEEDING

Schaperclaus (1933), classified the natural food of fish into three groups

1. "Main food" or the "natural food", which the fish prefers under favourable conditions and on which it thrives best.

2. "Occasional food" or the natural food that is well liked and consumed as and when available, and

3. "Emergency food", which is ingested when the preferred food items are not available on which the fish is just able to survive.

Certain microscopic planktonic crustacean groups and rotifers form the "mainfood" of spawn and fry (15 to 20 mm size range) of the Indian Major carps and majority of the other culturable species, with phytiplankton forming the "emergency food". Spawn and fry with a small and short straight intestine appear to digest rotifers and cladocerans fairly rapidly and thrive well on zooplankton. Phytoplanktonic algae are not so easily digested and, at least, some algal forms (Euglena, Oscillatoria, Microcystis, Filamentous green algae, etc.) remain undigested being ejected intact along with faecal matter. The food and feeding habits of the fingerlings of Indian Major carps differ markedly from those of their hatchlings and fry.

Recently Nikolskii (1963) divided food of fish into four categories according to the relationship between the fish and their food. These catagories are:

1. Basic food, which the fish usually consumes, comprising the main part of the gut contents.

2. Secondary food, which is frequently found in the guts of fish, but in small amounts.

3. Incidental food, which only rarely enters the gut and

4. Obligatory food which the fish consumes in the absence of basic food.

Adult fish, according to the character of diet they thrive on, have been classified into Herbivores, if they feed on vegetable matter; Carnivores, if their food comprises of animal matter; and Omnivores, if they subsist on mixed diet comprised of both vegetable as well as animal food. Nikolskii (1963) categorized fish acording to the extent of variation in the food consumed by them, such as

1. Euryphagic, feeding on a variety of foods

2. Stenophagic, feeding on a few selected types of food and

3. Monophagic, feeding on a single type of food.

The feeding behaviour is a species characteristic.

Cultured fish are often classified according to the trophic niche they occupy in a water-body. Following this system, fish have been grouped into

1. Plankton eating surface feeders. Ex. *Catla catla*

2. Column or Mid feeders. Ex. *Labeo rohita, Labeo bata, Labeo calbasu*

3. Bottom feeders. Ex. *Cirrhinus mrigala*

The food and feeding habits of the cultural species are given in Table 4.1:

TABLE 4:1. FOOD AND I EDING HABITS OF FISHES

Species		Food	Feeding Habit
Catla catla	Fingerlings -	Water fleas, Planktonic algae and some vegetable debris	Surface feeder
	Adult -	Crustaceans, algae, plants, rotifers, insects, vegetable debris	Planktophagus (Zooplankton main diet)
Labeo rohita	Fingerlings -	Vegetable debris, microscopic plants	
	Adult -	Vegetable debris, microscopic plants, decayed higher plants, detritus and mud	Predominantly column feeder
Cirrhinus mrigala	Fingerlings & Adults	Decayed plant and animal matter, algae, detritus, Planktophage mud, etc.	Bottom feeder . Omnivore
Cirrhinus cirrohosa	Fingerlings -	Vegetable debris, mud, detritus algae	
	Adults -	Detritus, sand and mud, decaying leaves	Bottom feeder, Herbivore
Wallago attu	Fry -	Water fleas, insects, fish fry	
	Fingerlings -	Insects, fish fry and fingerlings	Highly carnivorous and
	Adults -	Small and medium sized fish	predaceous
Channa striatus	Fry -	Water fleas, insects, fish fry	
	Fingerlings -	Dipteran larvae, Zooplankton, fish fry	Bottom feeder
	Adult -	Small fishes	Carnivore
Heteropneustus fossilis	Fingerlings -	Insect and worms	Bottom feeder
	Adults -	Insects, ostracods, debris and algae	Omnivore
Anabas testudineus	Fry -	Minute protozoans, animalcules and water fleas	
	Fingerlings -	Mosquito and other insect larvae, water fleas, etc.	
	Adult -	Insects, water-fleas, vegetable debris, fish, etc.	Mainly entomophage
Cyprinus Carpio	Fry -	Nauplii	Omnivorous
	Fingerlings -	Rotifers, Cyclops, Euglena, Oscillatoria	Omnivorous
	Adult -	Ostracods, insect including Chironomid larvac, Microcystis, Euglena, Clostredium	Omnivorous
Et. oplus suratensis	Fingerlings -	Animalcules and water-fleas, filamentous and unicellular algae	
	Adult -	Diatoms, filamentous algae, minute crustaceans, insects, leaves of aquatic macrophytes.	Omnivorous

BIOLOGY OF FEEDING IN FISH

Labeo rohita

This fish is common in the plains of North India. Rohu fry start feeding on plankton from the 5th day after hatching. Zooplankton and smaller phytoplankters were the main food items. The food consists of vegetable debris, animalcules, water fleas and sand or mud up to 20 mm length of the juveniles. From 20 mm length onwards rohu feeds on unicellular and filamentous algae also. The zooplanktonic organisms consumed include cladocerans, copepods, insect larvae, rotifers and nauplii. Among the crustaceans there is strong preference for Cyclops and among rotifers, Keratella is consumed heavily. In case of adults, phytoplankton is preferred over zooplankton.

The fish mainly feeds on mid surface waters although it explores other zones of the environment. The intensity of the feeding is high throughout in the juveniles. Feeding intensity of the adults is affected by maturation and spawning. The mature fishes showed considerable increase in the feeding intensity. Mature fishes show a slackening in feeding. Spent fishes again feed actively. This increase and decrease in feeding is more prominent in the females than the males. Males exhibit better feeding than females during spawning months and on the whole feeding is better in the males throughout the year than the females. However, the females feed more actively during post-spawning months than the males.

The age of rohu is determined from its scales. The scales of rohu showed growth rings in the form of carved out grooved rings which are found to be annular and hence suitable for age determination. Rohu was found to attain an average length of 310 mm, 500 mm, 650 mm, 800 mm, 850 mm, 890 mm, 920 mm, 940 mm, and 960 mm, at the end of the first to tenth year of the life. Thus the growth increment was 310 mm, 190 mm, 150 mm, 90 mm, 60 mm, 50 mm, 40 mm, 30 mm, 20 mm, and 20 mm, at the end of the first to the tenth year of life respectively.

Labeo rohita like other major carps, breeds naturally in the rivers and reservoirs and in the artificially constructed bundh type tanks where fertile conditions are stimulated during the spawning season. Large scale natural spawning occurs in flooded sections of rivers during the southwest monsoon months. Generally spawning grounds are located in the middle reaches of the rivers where flooded water inundates vast adjoining riparian lands. Breeding commences in the morning in rohu. The courtship is short-lived. The coiling of the two partners exerts pressure on the abundance of the mating pair, resulting in the expansion of ova and exudation of milt. All eggs are not laid at one time, but at intervals during which the pair keeps on moving. Fertilisation is external. The fertilised eggs are abondoned by the parents. A thick blanket of eggs are left behind on the spawning sites. The spent fish swim aimlessly for a while and commence their homeward journey along with receeding water. Heavy monsoon flood, capable of inundating vast shallow areas which form the breeding grounds of fish, stimulates spawning and is believed to be primary factor responsible for spawning.

Chanos chanos

This is commonly known as milk fish. It grows up to a length of 1500 mm. This is a marine and estuarine fish, suitable for cultivation in freshwater and brackish water ponds. It does not mature or breed in confined waters. It spawns in the sea near the coast. Eggs are pelagic, 12 mm in diameter. The embryos hatch out within 24 hrs. Larvae are 12-15 mm in length, occurring periodically in great quantities along sandy coasts in the estuaries. Fry and fingerlings are stocked in ponds and reservoirs. Fry reach a length 50 to 70 mm at the end of the first month and 120 to 150 mm at the end of second month. It reaches an average length of 40 cm and an average weight of 450 gm within a year. The fry and fingerlings feed on the phytoplankton belongs to the families Bacillariophyceae, Myxophyceae, etc. Adults feed on diatoms, copepods, larval bivalves, fish eggs, etc. In rearing ponds it feeds on the green bottom growth called "lab lab". This substance represents a biological complex consisting of decayed green algae, diatoms and detritus. When attains up to 450 gms in weight in tanks it feeds on a comparatively large portion of fresh filamentous algae and parts of higher plants. In freshwater ponds and tanks Chanos feeds on vegetable planktonic and epiphytic organisms and on decayed macro-vegetation.

BIOLOGY OF FEEDING IN PRAWNS

Penaeus monodon

This species occurs along both the coasts and more commonly along the northeast coast of India. It widely cultures in brackish waters. This species grows to a maximum of 340 mm. The major food items are higher plant matter, algae, diatoms, copepods, isopods, amphipods, decapods, gastropods, detritus, pieces of moulds, fish eggs, fish scales. The monthly growth rate of this species in natural waters is about 25-30 mm, whereas under culture conditions the observed growth is at a lower rate. In initial months the growth rate was less than 15 mm in length and 2.5 gm in weight per month, but in the later stages of culture, more than 28 mm in length and 10 gm in weight per month. Growth is fastest in the summer months than in winter months. Generally under pond conditions the species does not mature. As in other estuarine penaeid prawns, this species also shows catadromous migration, i.e., the post larval forms enter the estuaries and after a period of growth the young prawns return to the sea for breeding. Two peaks are observed in the incursion of post larval prawns into the estuary: November to January and April to June. The fertilised egg is released directly into the water and further development follows.

Macrobrachium rosenbergii

This species occurs in the freshwater zones of the rivers of the East and West coasts. Though a freshwater prawn, it migrates into the estuaries for breeding and spawns in areas where salinity fluctuates between 5 to 20 ppt. After the young grow into a size of 2 to 3 cms, they migrate up to the estuary to the freshwater habitats. They attain a length of 230 mm. The rate of growth ranges between 20 to 30 mm per month during different stages of growth. It is an omnivorous feeder, feeding on detritus, animal and vegetable matter. It attains maturity in the culture conditions and normally breeds in estuaries, It is suitable for cultivation in ponds and tanks.

5

LIFE HISTORY

EGG

Identification and segregation of different species from a mixed collection of fish seed are of vital importance for selective stocking of ponds in order to avoid wasteful culture of economic species.

IDENTIFICATION OF EGGS OF SOME CULTURED FISH BASED ON THEIR CHRACTERISTICS

Nature	*Diameter (mm)*		*Colour*	*Species*
Non-floating				
a. Non-adhesive	5.3 -6.5	Round	Yolk light-red	*Catla catla*
	5.5	Round	Brownish	*Cirrhinus mrigala*
	5.0	Round	Reddish	*Labeo rohita*
	3-4	Round	Bluish	*Labeo calbasu*
	4.5	Round	Bluish	*Labeo gonius*
b. Adhesive				
Non filamentous	1.3 - 1.5	Oval	Greenish	*Clarias batrachus*
	1.5	Round	Greenish	*Heteropneustes fossilis*
	1.2-1.5	Round	Yellowish	*Wallago attu*
Floating	1.0	Round	Golden amber	*Channa punctatus*
	1.25-1.35	Round	Amber	*Channa striatus*
	0.7	Round	Transparent	*Anabas testudineus*

LARVAE

IDENTIFICATION OF LARVAL STAGES OF FISH FAMILIES

Prolarval Stages

1.	Yolksac round in shape	Odontaspididae and Lamnidae
2.	Yolksac oval in shape	Torpedinidae and Rajidae
3.	Yolksac oblong in shape	Rhinobatidae
4.	Round yolksac without oil globules	Scyliorhinidae
5.	Round yolksac with oil globules	Carchahinidae and Sphyrinidae
6.	Unsegmented yolk, barbels absent Anal opening near about the middle of the body	Gobiidae
7.	Segmented yolk, barbels present	Squalinidae
8.	Unsegmented yolk, barbels absent	Pristidae
9.	Anal opening situated in the posterior 1/3 of the body	Clupeidae
10.	Anal opening near about the middle of the body	Anabantidae
11.	Barbels absent	Channidae
12.	Yolk sac posteriorly elongated	Dasyatidae
13.	Yolksac not elongated posteriorly	Myliobatidae and Mobulidae

Postlarval Stages

Cyprinidae : Eyes pigmented, air bladder conspicuously dark, embryonic fin-fold continuous, anal opening in the posterior 1/3 of the body, mouth formed, pectoral fins enlarged, barbels absent or present.

Siluridae : Eyes pigmented and small, 1-4 pairs of well developed barbels, air bladder absent, embryonic fin fold continuous, anal opening near about the middle of the body, mouth formed.

Channidae : Eyes pigmented, pectoral fin well developed, air bladder present, embryonic fin fold continuous, barbels are absent.

Cichlidae : Eyes large and pigmented, barbels absent, anal situated in the anterior half of the body, embryonic fin fold is continuous and narrow.

Mugilidae : Barbels absent, first dorsal fin indicated as a narrow strip of the embryonic fin fold, second dorsal well differentiated, anal also differentiated, ventral fin rudimentary, anal opening near the middle of the body.

SPAWN AND FRY

Diagnostic Characters of the Spawn and Fry of *Cirrhinus mrigala*

Hatching

1. Average size is 4.2 mm.
2. Yolk more or less club shaped.
3. 28 pre-anal and 14 post-anal myotomes present.

24 hrs after hatching

1. Average size 6.21 mm.
2. Narrow end and yolk does not end in a point sharply.

36 hrs after hatching

1. Average size is 6.70 mm.
2. Anterior profile of yolk sac more or less straight.

48 hrs after hatching

1. Average size is 6.84 mm.
2. Anterior profile of yolk sac more or less straight.

72 hrs after hatching

1. Average size is 7.2 mm.
2. Few black chromatophores in the caudal region.

96 hrs after hatching

1. Average size is 7.38 mm.
2. Black chromatophores on the dorsal fin rudiment, towards anterior region.
3. Black chromatophores below the tip of the notochord, confined to a semicircular area.
4. Dorsal fin seprating from the embryonic fin fold.

5th day after hatching

1. Average size is 9.0 mm.
2. Caudal rays about 7, not branching.

Hatchling

Larva—24 hr after hatching

Larva—36 hr after hatching

Larva—48 hr after hatching

Larva—72 hr after hatching

Post larva—4th day after hatching

Post larva—5th day after hatching

Post larva—6th day after hatching

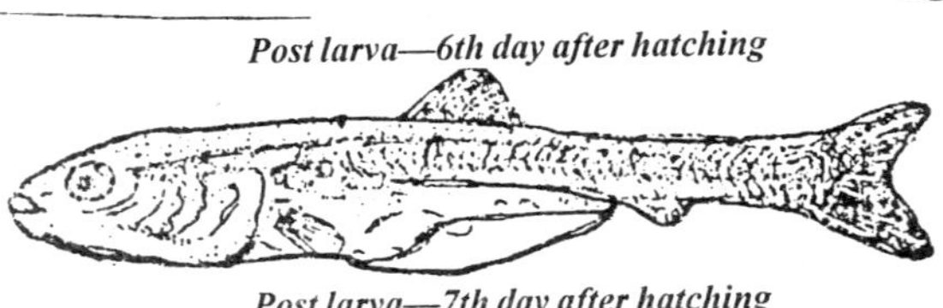

Post larva—7th day after hatching

Post larva—10th day after hatching

Fig. 5.1 Larval developmental stages of Cirrhinus mrigala

6th day after hatching

1. Average size is 11.5 mm.
2. Dorsal half of the body is in yellowish green.
3. Two faint somewhat crescentric markings on caudal peduncle.
4. Anal fin bud present.

7th day after hatching

1. Average size is 13.2 mm.
2. Black chromatophores form a triangle anterior to the origin of the caudal peduncle.
3. Dorsal fin separated from embryonic fold.

10th day after hatching

1. Average size is 15.6 mm.
2. Black chromatophores absent in dorsal fin margin.

15th day after hatching

1. Average size is 27 mm.
2. Body is more pigmented.

Diagnostic Characters of the Spawn and Fry of *Catla catla*

Hatching

1. Average size is 4.68 mm.
2. Bulbus part of the yolk and narrow part are equal in length.
3. 26 pre-anal and 14 post-anal myotomes present.

24 hrs after hatching

1. Average size 5.80 mm.
2. Narrow end and yolk does not end in a point sharply.

36 hrs after hatching

1. Average size is 6.40 mm.
2. Anterior profile of yolk sac convex.

48 hrs after hatching

1. Average size is 6.48 mm.
2. Anterior profile of yolk sac convex.

72 hrs after hatching

1. Average size is 7.3 mm.
2. Dark somewhat triangular spot in caudal peduncle.

96 hrs after hatching

1. Average size is 7.56 mm.
2. Semicircular mark, formed of black chromatophores, seen on ventral side of the caudal towards end of the notochord.
3. Lip margins are very thick.
4. Dorsal fin separating from the embryonic fin fold.

5th day after hatching

1. Average size is 9.0 mm.
2. Caudal rays about 18, branching distally.

6th day after hatching

1. Average size is 11 mm.
2. Body pale yellow.
3. Anal fin bud present.

7th day after hatching

1. Average size is 12 mm.
2. Dorsal fin separated from embryonic fold.

10th day after hatching

1. Average size is 16.5 mm.
2. Lips are thick.
3. Mouth somewhat upturned.

15th day after hatching

1. Average size is 23 mm.
2. Margin of the dorsal fin dark.

Life History, Eggs and Larval Stages of *Labeo rohita*

Rohu attains maturity towards the end of the second year of life. It spawns only once a year and the spawning duration is short. The spawning season of rohu generally coincides with the southwest monsoon.

Fertilised eggs are round, transparent, demersal, nonadhesive and red in colour. The yolk sphere contains no oil globule. Fully fertilized eggs are 5 mm in diameter. The newly hatched larva measures 3.5 - 4.5 mm. It has a transparent, laterally compressed body and is characterised by the presence of gill slits, pectoral fin and median fin fold.

Post larvae, just after absorption of yolk measure 6.5 mm to 7.5 mm. Rudiments of dorsal, anal and pelvic fins appear on 30th, barbels on the 21st day and liver on the day after hatching. The airbladder originates 2 1/2 days after hatching.

After 24 hours of hatching

1. Average size of larva 5.5 mm.
2. A few chromatophores are seen on the head above the eyes.
3. Embryo pale yellow in colour.
4. Yolksac sharply ending distally to a point.
5. Striations seen in the caudal.
6. Notochord turned upwards only at the very end.

After 36 hours of hatching

1. Average size about 5.9 mm.
2. Lower lip clear.
3. Anterior profile of yolk sac more or less straight.
4. A few black chromatophores on dorsal margin of the yolk sac throughout its length, a few on the dorsal fin. No chromatophores on the caudal region.
5. Notochord slightly upturned at the tip.
6. Mouth appears as a slit.

After 48 hours of hatching

1. Average size 6.2 mm.
2. Yolk sac convex anteriorly.
3. Embryo yellow in colour.

4. Air bladder distinct.
5. Pectoral fin prominent.
6. No chromatophores on the ventral side of the air bladder.
7. Head dark and the body is faint yellow.

After 72 hours of hatching

1. Average size 7 mm.
2. Pale yellow in colour.
3. Black chromatophores on the head with a few in between the eyes.

The charactistics of rohu post larvae after three days of hatching.

4th day after hatching

1. Average size 7.6 mm in length.
2. Yolk sac completely absorbed on the 4th day after hatching.
3. Distinct black chromatophores seen behind the eyes on the head region.
4. Notochord bent at the tip.
5. Commencement of caudal rays.

6th day after hatching

1. Average size 10 - 10.5 mm in length.
2. Lower jaw bigger and upturned.
3. Nostrils are prominent.
4. Dorsal fin with 9 rays, anal with faint rays, pelvic fin with bud seen.
5. Air bladder divided into 2, the anterior part globular and posterior in the form of elongated triangle.
6. Dorsal part of the body is yellow.
7. Caudal rays 22.

7th day after hatching

1. Average size 11 mm in length.
2. Lips are very thick.
3. Notochord sharply bending upward.
4. Ventral embryonic fin fold, stretching from abdominal region up to anus.

5. Dorsal embryonic fin fold visible as far forward as opposite to the anus.
6. Two black chromatophore clusters at the peduncle and two crescent shaped patches on the caudal fin.
7. Caudal fin with 22 rays.
8. Caudal fin less deeply forked.
9. Pelvic fin with 2-3 rays. Dorsal fin with 2-11 rays.

8th day after hatching

1. Average length is 12.5 mm.
2. Dorsal fin with 14 rays sparsely distributed yellow pigment near base on the rays.
3. Anal fin with 7 rays.
4. Ventral embryonic fold ending in the anal region.
5. Caudal fin with 22 rays.
6. Pelvic fin with 5 rays.
7. Body golden yellow, dorsal half more predominantly.
8. Prominent black chromatophores on the head.
9. Black chromatophores scattered all over the body exhibiting no pattern.
10. The dark crescentric areas at the base of the caudal fin and a few black chromatophores on the embryonic fold connecting the caudal fin with the anal fin.

12th day after hatching

1. Average length is 19 mm.
2. Maxillary barbels present.
3. Body golden coloured on the dorsal side and dirty yellow ventrally.
4. Dorsal fin with 3-13 rays and 1 fin with 7 rays, except for the anterior one, the rest are branched.
5. Ventral fin with 7 rays, except two anterior ones, the rest are branched.
6. Caudal fin rays 34, a broad triangular black band on the caudal peduncle across its entire with the apex facing the head.

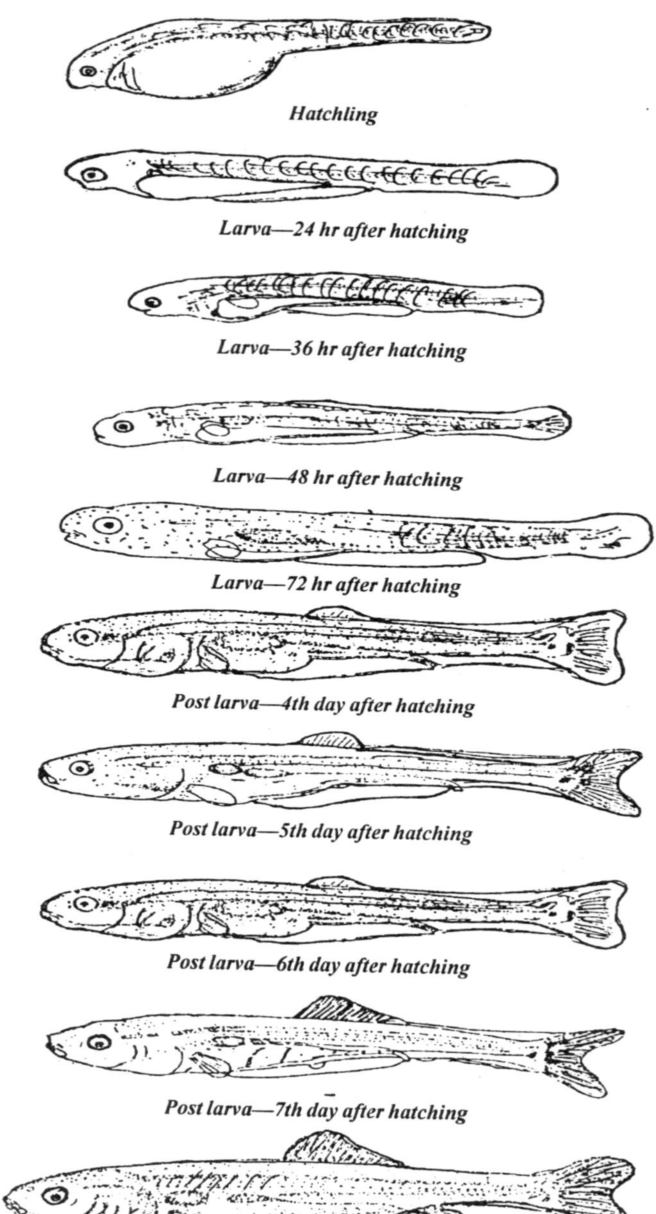

Fig. 5.2 Larval developmental stages of Catla catla

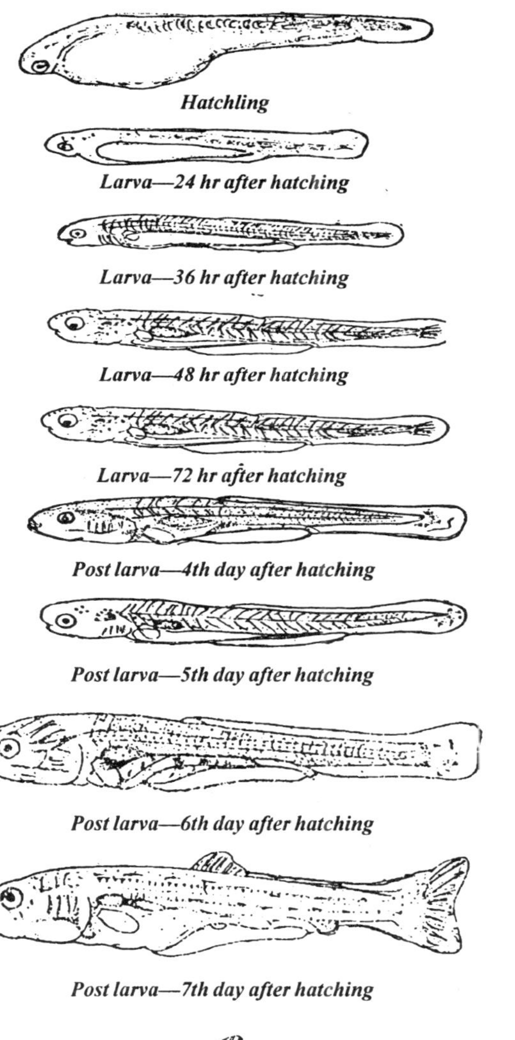

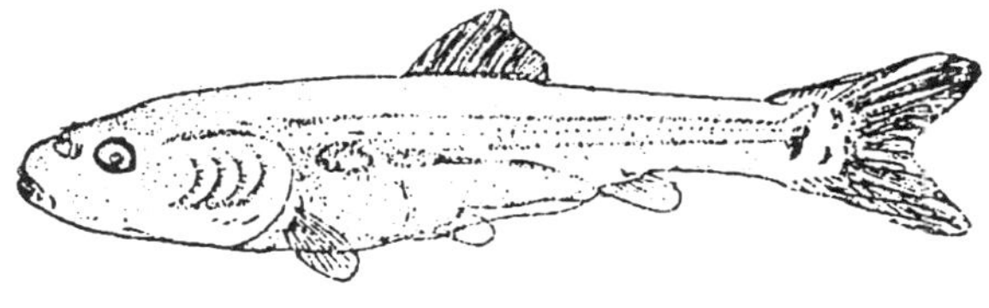

Post larva—10th day after hatching

Fig. 5.3 Larval developmental stages of Labeo rohita

15th day after hatching

1. Average length is 23 mm.
2. Anal fin with 2-5 rays.
3. Caudal rays 36.
4. The origin of caudal rays and dividing the fin into two.

18th day after hatching

1. Average length is 25 mm.
2. Body colour is dirty yellow above lateral line and yellowish white below it.
3. Black chromatophores on the anterior margin of the dorsal fin in the form of dotted line.
4. Ventral fin with 9 rays.
5. 32 caudal rays.
6. Two prominent greyish crescents, one on each half of the caudal fin lobes behind the dark triangular areas on the caudal peduncle.

20th day after hatching

1. Average length is 26 mm.
2. Body is golden coloured.
3. Scales are prominent and fully covering the body.
4. Two dark bands one on above and one below the lateral line along the body up to the caudal region.
5. Dorsal fin with 3-13 rays.
6. Caudal rays 34.
7. All the rays covered with orange pigment dots.
8. A black band along the entire width of the caudal peduncle concave posteriorly and irregular in shape along the anterior margin. Behind this band two faint crescentric areas separated by two colourless streaks.

25th day after hatching

1. Average length is 30 mm.
2. Body is golden coloured with coppery reflection from operculum.
3. Barbels prominent.
4. Upper margin of the eye faintly orange.
5. Upper lobe of caudal fin larger and pointed compared to somewhat round edge of smaller lower lobe.

Life History, Eggs and Larval Stages of *Channa punctatus*

The fertilised eggs are round, non-adhesive, buoyant, free and straw yellow in colour. The yolk is capped by a distinct blastodisc and contains an oil globule liable to distingration to smaller 5 to 6 globules, unequal in size. Perivitalline space is of medium width. The eggs are 1.2 mm in diameter and that of yolk is 0.9 mm.

The sequential development after fertilisation is as follows:

10 hrs after fertilisation

1. Anterior axis is distinguishable.
2. Cephali portion is very much broader.
3. Fore brain region discernable.
4. Invasion of the yolk complete.

14 hrs after fertilisation

1. The embryo with somites is embeded all over its length in the yolk mass.
2. Optic cups are very clear.
3. The general height of the embryo will be around 1.0 mm.

16 hrs after fertilisation

1. The embryo shows 9 mesodermal somites present.
2. Notochord is more discrete and the fore, mid and hind brain regions are also defined.
3. Cephalic region is further broadened.

17 hrs after fertlisation

1. The somites range from 13 to 15 in number.
2. Ectodermal thickning to form lens of the eye is indicated.
3. Tip of the tail is from the yolk at the vast somite.
4. Embryonic fin fold appears.
5. Kupter's vesicle is clearly visible.
6. Cephalic region is depressed.
7. The general height of the embryo is around 1.1 mm.

20 hrs after fertilisation

1. The embryo possess 22 somites.
2. The embryonic fin fold on the ventral side extends up to 10th somite.
3. The lens is now fully formed in the eye.
4. Olfactory placode indicated.
5. Kupter's vesicle gets diminished.
6. Blood circulation can be seen over to yolk into the rudimentary heart lying anterior to the yolk sac.
7. The heart beats very rapidly.

22 hrs after fertilisation

1. The embryo possess 24 somites.
2. The embryo encircles the yolk and covers practically the entire capsule.
3. The tail is relatively free.
4. Olfactory pits and auditory vesicles are now prounced.
5. Some melanophores appear above the neural chord, over the trunk and caudal regions.
6. Kupter's vesicle disappears.
7. Heart beat is very rapid.

24 hrs after fertilisation

1. Twitching movement of the embryo is vigorous and lashes the tail vigorously against the capsule, ther by springing up the head which supresses the wall and brings the embryo out.
2. The newly hatched larvae are dull brown in colour and are generally 2.7 mm in length and 1.1 mm in height.
3. The fin fold originates dorsally at the second myotome, runs ventrally the region of the vent and goes up to the yolk sac.
4. The third and fourth ventricles are very prominent.
5. Linear melanophores appear prominently on the lateral side of the body.

3 hrs after hatching

1. The average length of the embryo will be 3.1 mm and the height is 1.4 mm.

2. The heart is two chambered. Circulation can be seen around the notochord in addition to the brain and the yolk.
3. Blood corpuscles are reddish yellow in colour showing formation of haemoglobin.
4. The anal opening is marked by a clear slight invagination.

8 hrs after hatching

1. The length is 3.5 mm and the height is 1.1 mm.
2. The myotomes are 31 in number.
3. Some melanophores appear on the ventral side of the notochord and the dorsal side of the body.
4. Circulation is conspicuous at the optic region.
5. Some pigments are interconnected along the intermyoformal septa.

15 hrs after hatching

1. The embryo length will be generally 4.2 mm and the height is 10 mm.
2. The auditory capsules moves closer to the eye.
3. Pigmentation over the iris became denser.
4. Mesenteron is seen as a blind tube and nephridial duct are conspicuous.
5. The location of the pectoral fin is clearly marked by a thickened conical patch.

24 hrs after hatching

1. The length of the embryo will be 4.3 mm and the height is 10 mm.
2. Myotome number will be 32.
3. Buccal invagination appeared.
4. The air sac is differentiated as small tube below the pectoral fin bud.
5. Eye is fully pigmented. Pectoral bud is well prounced.

2 days after hatching

1. The length of the larva will be 4.6 mm.
2. The pectoral fin is paddle shaped with undulating dorsal margin.
3. Larvae move horizontally in schools.

4. The lower jaw is well developed.

3 days after hatching

1. The length will be 5.0 mm.
2. Pectorals which are vigorously, have been vascularised with a distinct semicircular vessel running across them.
3. Opercula are membranous.
4. Yolk is fully absorbed.

6 days after hatching

1. The length is 5.2 mm.
2. Yellow pigments are seen on the dorsal side giving a band like appearance. A similar yellow band is seen on the lateral aspect.

8 days after hatching

1. The length will be 5.3 mm.
2. Rudiments of two caudal fin rays are seen in the caudal.

11 days after hatching

1. The length will be 6.1 mm.
2. Colour bands are more prominent.
3. Caudal fin rays 5 in number and purals are indicated as basal thickenings.

18 days after hatching

1. The length will be 6.3 mm.
2. 8 caudal rays can be distinguished, the middle ones showing articulation.

20 days after hatching

1. The length will be 7.3 mm.
2. Now the fry shows nearly all the adult characteristics except the colour patterns.
3. Fin rays formation generally will start from 18th day onwards.
4. There are 14 rays in the caudal, 19 in the anal, 25 in the dorsal and 17 in the pectoral at this stage. Aerial respiration is noticed even from this stage.

The ventral fins differentiated on the 23rd day when the fry measure 14 mm. Scales appear on the body by the 26th day when they are 18 mm long.

DIAGNOSTIC CHARACTERS OF CARP FRY (14-25 MM)

Major carps : Number of undivided dorsal fin rays > 11

Minor carps : Number of undivided dorsal fin rays 11 or < 11

Catla catla : Large head. Dorsal profile convex and ventral profile concave. Margins of the dorsal and caudal fins are thicker. Ist ray of the dorsal fin black. No barbels.

Labeo rohita :A dark diffused transverse band present at the caudal peduncle. Lips are fringed.

Cirrhinus mrigala : Small head and slender body. A more or less triangular dark spot at the caudal peduncle. No barbels are visible.

DIAGNOSTIC CHARACTERS OF ADVANCED FRY AND EARLY FINGERLINGS OF CARPS (30 - 100 MM):

Catla catla : Large head. No distinct spot on the body or at the caudal peduncle. No barbels. Lips thick but not fringed.

Labeo rohita : Dark band at the caudal peduncle persists. Lips fringed. Maxillary barbels are prominent. A pair of rostral barbels also appears.

Cirrhinus mrigala : The spots on the caudal peduncle becomes diamond shaped. Barbels are faintly visible. Lips are not continuous. A few longitunidal lines are visible on the body due to the pigments on the scales.

DEVELOPMENTAL STAGES OF *PENAEUS MONODON*

The gaint tiger prawn, *Penaeus monodon,* was reared at a temperature of 26.5 to 28.5 ° C and the salinity of the water was 30.2 to 33.5 ppt.

Eggs

The eggs are small with a narrow perivitelline space. The egg diameter varied from 0.25 to 0.27 mm and the yolk mass 0.22 to 0.24 mm.

The developing nauplius fills up the entire space inside the egg. The eggs hatched out 1-17 hours after spawning.

The following abbrevations are used :

MTL : Mean total length

MW : Mean width

MCL : Mean carapace length

MFS : Mean length of the longest pair of furcal setae.

Nauplius I

MTL : 0.31 mm (0.29 - 0.32 mm)

MW : 0.17 mm (0.17 - 0.18 mm)

MFS : 0.12 mm (0.10 - 0.13 mm)

1. Furcal setae 1 + 1; minute posterodorsal tooth present.
2. All the setae are non-plumose.
3. Antennule with 2 inner lateral setae distal one longer than proximal.
4. Antenna exopod bears 5 long setae along inner and distal margin.
5. Mandible with 3 long distal on exo and endopods.
6. The duration of this substage was 3-4 hrs.

Nauplius II

MTL : 0.32 mm (0.31 - 0.32 mm)

MW : 0.10 mm (0.17 - 0.18 mm)

MFS : 0.14 mm

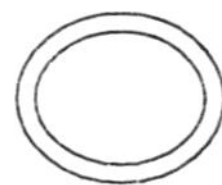

Fig. 5.4 Egg of Penaeus monodon

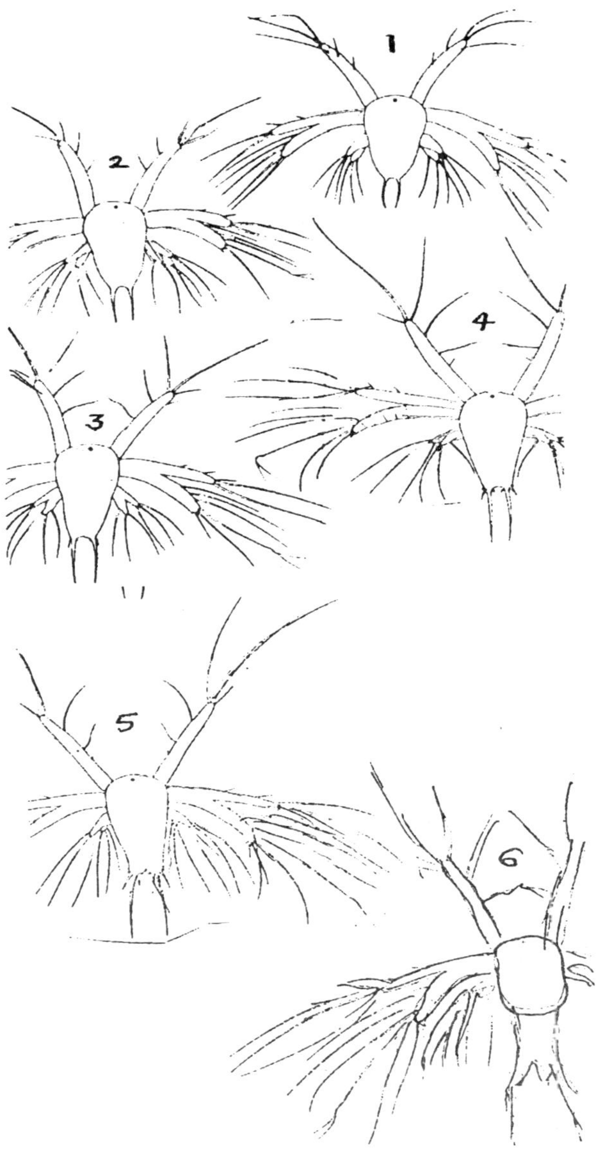

Fig. 5.5 Larval stages of Penaeus monodon

1. Nauplius I, 2. Nauplius II, 3. Nauplius III
4. Nauplius IV, 5. Nauplius V, 6. Nauplius VI

1. Furcal setae are 1 + 1, posterodorsal tooth absent.
2. Setae are plumose.
3. The duration of the substage was 3-4 hrs.

Nauplius III

MTL : 0.33 mm (0.31 - 0.34 mm)

MW : 0.17 mm (0.17 - 0.18 mm)

MFS : 0.18 mm (0.17 - 0.20 mm)

1. Furcal setae 3 + 3.
2. Antenna exopod with 6 plumose setae and a setal rudiment.
3. The duration of the substage was 4-5 hrs.

Nauplius IV

MTL : 0.36 mm (0.34 - 0.36 mm)

MW : 0.17 mm (0.17 - 0.18 mm)

MFS : 0.18 mm (0.17 - 0.20 mm)

1. Furcal setae 4 + 4.
2. Frontal organs seen.
3. Antenna exopod with faint segmentation.
4. Base of the mandible swollen.
5. The duration of the substage was 5-6 hrs.

Nauplius V

MTL : 0.39 mm (0.36 - 0.41 mm)

MW : 0.18 mm

MFS : 0.25 mm (0.22 - 0.25 mm)

1. Furcal setae 6 + 6.
2. Antennule with minute seta added outer lateral margin opposite to origin of distal inner lateral seta, faint segmentation seen in proximal half.
3. Antenna exopod with outermost setae but still non-plumose.
4. The duration of the substage was 10-12 hrs.

Nauplius VI

MTL : 0.49 mm (0.46 - 0.53 mm)

MW : 0.20 mm (0.18 - 0.20 mm)

MFS : 0.30 mm (0.28 - 0.32 mm)

1. Furcal setae 7 + 7.
2. Frontal organ prominent.
3. Carapace rudiment seen.
4. The duration of the substage was 15-24 hrs.

Protozoea I

MTL : 1.06 mm (1.05 - 1.09 mm)

MCL : 0.47 mm (0.44 - 0.49 mm)

1. Frontal organs rounded.
2. Telson with 7 + 7 setae.
3. Antennule with 3 main segments.
4. Antenna with 9-10 segmented exopod bearing 11 setae on inner and distal margin.
5. First and Second maxillipedes are well developed, whereas Third maxillipede absent.
6. The duration of this substage was 36 - 48 hrs.

Protozoea II

MTL : 1.06 mm (1.65 - 1.68 mm)

MCL : 0.72 mm (0.70 - 0.77 mm)

1. Telson with 7 pairs of furcal setae.
2. No change in the First and Second maxillipeds. Third maxilliped was absent.
3. The duration of this substage was 36 - 48 hrs.

Protozoea III

MTL : 2.28 mm (2.14 - 2.38 mm)

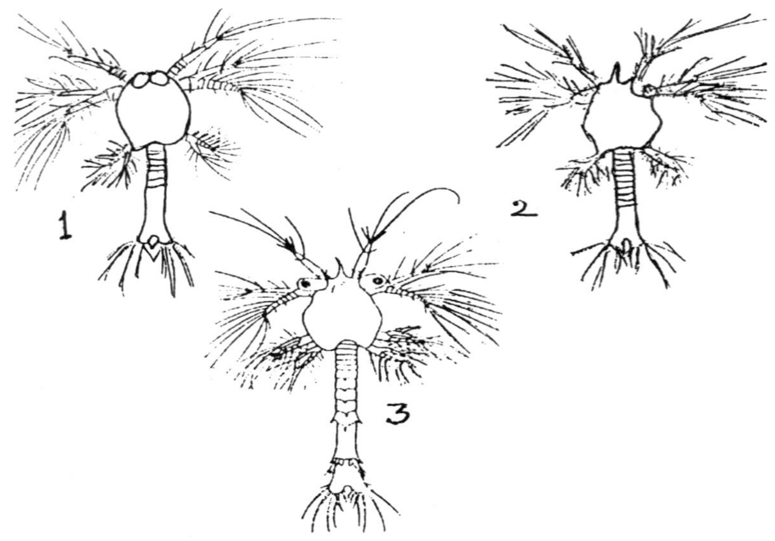

Fig. 5.6 Larval stages of ***Penaeus monodon*** ***1. Protozoea I, 2. Protozoea II, 3. Protozoea III.***

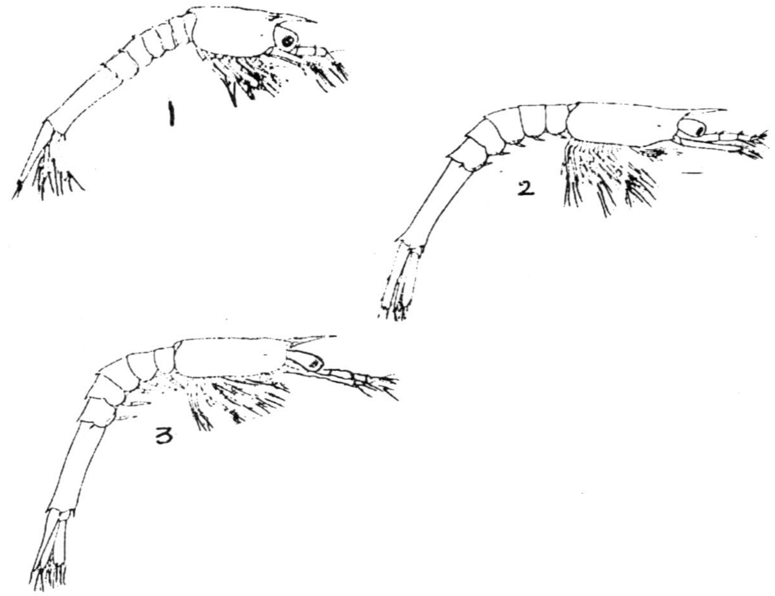

Fig. 5.7 Larval stages of ***Penaeus monodon*** ***1. Mysis I, 2. Mysis II, 3.Mysis III.***

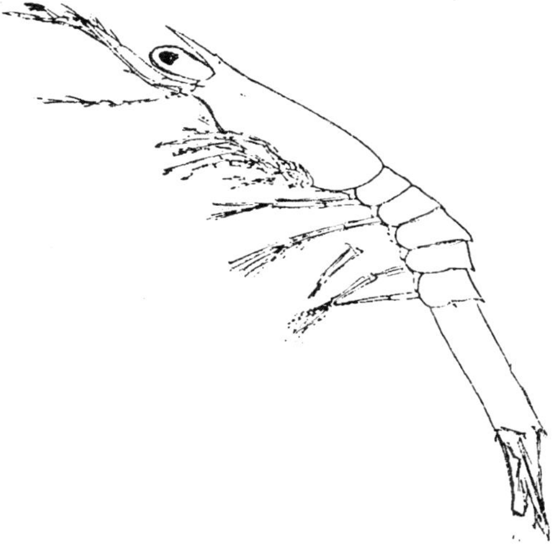

Fig. 5.8 Intermediate stage of ***Penaeus monodon***

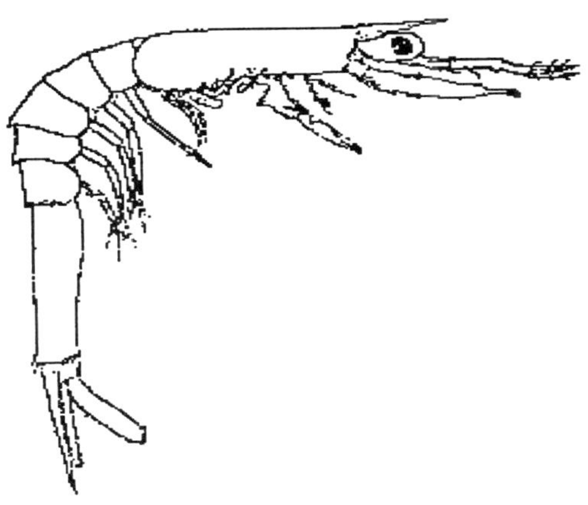

Fig. 5.9 Post larva of ***Penaeus monodon***

MCL : 0.79 mm (0.70 - 0.84 mm)

1. Rostrum long.
2. Telson with 8 pairs of furcal setae.
3. Third maxillipede was developed as biramous bud.
4. The duration of this substage was 36-48 hrs.

Mysis I

MTL : 3.79 mm (3.65 - 3.96 mm)

MCL : 1.18 mm (1.14 - 1.23 mm)

1. Carapace with rostrum longer than eye.
2. No rostral tooth.
3. Supraorbital, pterygostomial and hepatic spines present.
4. Telson with 8 + 8 short stout setae.
5. Antennule I is 3 segmented.
6. Antennule II expod unsegmented.
7. The duration of this substage was 24 - 48 hrs.

Mysis II

MTL : 4.16 mm (3.90 - 4.37 mm)

MCL : 1.39 mm (1.34 - 1.47 mm)

1. Rostrum usually without teeth.
2. The duration of this substage was 24 -48 hrs.

Mysis III

MTL : 4.24 mm (4.00 - 4.59 mm)

MCL : 1.50 mm (1.48 - 1.53 mm)

1. Rostrum with 1 tooth.
2. The duration of the substage was 24 - 48 hrs.

Intermediate Stage

1. Rostrum with 1 distinct tooth.

2. The duration of this substage was 24 - 30 hrs.

Post Larva 1

MTL : 4.56 mm (4.45 - 4.70 mm)

MCL : 1.46 mm (1.40 - 1.54 mm)

1. Rostrum with 1 distinct tooth.

6

SEED

STOCKING MATERIAL

PROCUREMENT OF STOCKING MATERIAL

Chaudhari (1967) classified the important culture fishes, according to their breeding habits, into three distinct catagories.

1. Fishes which breed in ponds

   Ex: *Cyprinus carpio*
   Tilapia mossambica
   Etroplus suratensis
   Channa species

2. Fishes which do not ordinarily spawn in ponds but breed in flooded rivers and adjoining areas

   Ex: *Catla catla*
   Labeo rohita
   Cirrhinus mrigala
   Labeo calbasu
   Hypophthalamichthys molotrix
   Ctenopharyngodon idella

3. Fishes which mainly inhabit the sea or brackish water and which neither

spawn in ponds nor could be induced so far to spawn successfully therein

Ex: *Chanos chanos*
Lates calcifer
Mugil species

THE SOURCE OF STOCKING MATERIAL

The stocking material required in various parts of the country for the fish culture operations comprises the eggs, spawn, fry and fingerlings of Indian major carps. The stock is procured from natural source by

1. Collecting eggs from breeding grounds
2. Collecting spawn, fry and fingerlings from rivers
3. Breeding fish in bundh type tanks and
4. Breeding fish through hypophysation technique

Normally carp families breed in rivers during monsoon months under the natural conditions. During collection along with fish seed of economic importance, uneconomic species also be collected. Traditional methods of collection of carp spawn and fry from natural resources like rivers, bundhs are still in vogue. But lack of scientific knowledge and awareness about the actual method of seed collection, spawn collection and usage of proper nets and other seed transport mechanisms causing heavy loss to the economically important seed of the fishes.

RIVERINE CARP SPAWN COLLECTION

Indian major carps catla (*Catla catla*), rohu (*Labeo rohita*) and mrigal *(Cirrhina mrigala)* are highly preferred fish species for cultivation in freshwater ponds and tanks throughout India. All the Indian major carps inhabit rivers and the original spawning grounds are the flooded rivers. Traditional methods are applied for the collection of spawn and fry of carps. Inadequate scientific knowledge with reference to method of collection and transport mechanisms causing heavy loss of carp seed. Though the carp seed is available in plenty, but lack of scientific approach causing havoc to the fishing industry.

RIVERINE FISH SEED

From all the five rivers systems of India, namely, the Ganga, Brahmaputra, Indus river system in the north, Peninsular east coast and

West coast river system in the south, the fish seed are being collected in huge quantities. Ganga river system alone contribute about 89% of the fish seed collected from natural resource in the country.

BUNDH BREEDING

Bundhs are special type of perennial and seasonal tanks or special tanks where riverine conditions are simulated during monsoon months. Bundhs are of two types, viz., Perennial bundhs, often designated as wet bundhs, hold water throughout the year. But the other ones dry bundhs, tanks which retain only during the monsoon season. Bundhs provide large shallow marginal areas which serve as spawning grounds for the fish. In case of wet bundh, a perennial pond, the major portion of bundh gets submerged with water, the excess flowing out through the outlet known as Bulans. The shallow areas of bundhs, where fish actually spawn, are called Moans. The outlet is protected by a bamboo fencing known as Chhera. The flow of water through the outlet can be controlled by blocking the Chhera with straw and mud.

The dry bundh, which is a shallow depression enclosed on three sides by an earthern embankment, which impounds fresh rain water from the vast catchment area during the monsoon season. In dry bundhs, a selected number of major carp breeders in the ratio of 1 female to 2 males (1:1 by weight) are introduced. At first, smaller sized mature fish get stimulated to breed and in order to spawn, migrate either to the shallow areas of the bundh itself or to those adjoining it.

SPAWNING

Spawning in both wet and dry bundhs usually occur after heavy showers when large quantities of water rushes into the bundh. In wet bundh, during summer the deeper portion of the pond retains the water containing major carp breeders. After heavy showers, freshwater from catchment area rushes into the bundh in the form of streamlets. The major portion of bundh gets submerged with water, excess flowing out through outlet which is protected by bamboo fencing made up of straw and mud. The spawning occurs in the shallow region of the bundh and the conditions required for natural breeding of carps in wet bundhs are not clearly known. Dry bundhs are considered to be one of the most reliable means for mass breeding of grass carp and silver carp. Spawning may occur at night and during the bright sun in the forenoon. Mrigal and rohu occupies marginal area and breed in the morning. Catla may occupy deeper water due to its big size and breed from noon till evening. In major carps fertilisation is external. The fertilised eggs are left behind by the parents. The eggs get drifted to the edges of the bundh or get washed down. In bundhs a large number of eggs settle down in shallow water giving a pale whitish appearance. But the adult fish move towards the deeper areas.

No single factor can probably be attributed to spawning of major carps in rivers and bundh type tanks. The temperature of water at spawning grounds has been reported to range from 22 to 33° C in various environments.

COLLECTION OF EGGS AND SPAWN

Soon after the spawning process is over, the eggs are collected from the bundh with the help of nets made of mosquito netting cloth (Gamaha). The eggs collected are released for hatching either in improvised pits or in double walled hatching hapas or in cement hatcheries. The hatching pits of the size of 448 × 244 × 46 cm are excavated on the bank with arrangement for the supply of water. Each pit of this size may contain about 900,000 to 2,200,000 eggs of which 2.5% to 25% hatch successfully. A double walled hapa, which is fixed in the bundh itself, consists of an outer hapa measuring 182 × 91 × 91 cm and the inner one measuring 152 × 76 × 46 cm. The former is made up of ordinary cloth and the latter of round meshed mosquito netting. These hapas are laid out in perfect horizontal positions, one within the other, the inner having a water depth of 23 to 30 cm. The hatchlings passing out into the outer hapa are retained for three days for conditioning before stocking. The survival rate ranges from 32 to 55%.

Spawn collected from the rivers and wet bundhs, besides comprising a mixture of numerous species, usually has large sized associates, many of which are predaceous. The spawn is, therefore, sieved through a round meshed mosquito netting cloth on to a muslin cloth suspanded in water in order to seperate the different species.

INDUCED BREEDING IN FISH

The Indian major carps, Catla, Rohu, Mrigal and Chinese carps breed in the rivers during monsoon season. The Indian major carps breed in the specialised environments, wet and dry bundhs, wherein fluctuating conditions are stimulated during monsoon months. So, the fish culturists have to depend on for the fish seed only from river ecosystem. Thus the seed collected from rivers consist not only of preferred species and also may include uneconomic and even predatory forms. Further, as such seed are available at some specific centres located on rivers only. Great difficulty is encountered in transporting them to markets. Such a process, however, besides involving heavy mortality during transit, is expensive. The limitations and uncertainties in procuring the seed, prompted the need for developing suitable methods to obtain the pure seed of cultivated Indian and Chinese carps. The present day concept of the role of pituitary in the vertebrate reproduction, may be said to have originated from the experiments of Aschheim and Zondek in 1927, who noticed that pituitary implants accelerated sexual development in mice. However, the concept of the application of pituitary injections for successful spawning of fish is due to Houssay (1930) of Argentina. Brazil was the first country to develop a technique for hypophysation. Von Ihering and his co-workers conducted experiments with various hormone injections on the lines of Houssay (1930) and achieved success in 1934. Since then, Brazilian fish culturists have been employing this technique to obtain seed from indigenous fishes as a part of their routine piscicultural programmes. Major breakthrough in hypophysation research in India came to the limelight in 1957, when success was achieved in breeding of Indian major carps, by Chaudhuri and Alikunhi with carp pituitary extracts. Since then, the employment of this technique has spread widely and now forms a regular part of piscicultural programmes in many parts of India. Chinese carps have also been successfully bred in 1962 by employing similar techniques. The technique of spawning the fishes under controlled conditions by administering pituitary hormones is known as Induced Breeding or Hypophysation.

PITUITARY GLAND

In fish, pituitary gland is an important endocrine gland. It is very small, round in shape located on the ventral side of the brain immediately behind the optic chiasma in a concavity on the floor of the brain box known as sella tunica. In higher vertebrates, the pituitary gland secretes large number of hormones, some of them induces gonad stimulating hormones—gonadotropin, growth stimulating hormone- somatotropin and the thyrotropic hormone—thyroxin. In certain Teleost fishes pituitary secretions include follicle stimulating hormone (FSH) and leutinising hormone (LH) like gonodotropins. The synthesis and secretions of gonadotropins is influenced by several environmental factors like temperature, photoperiod, monsoon rains, etc. All the gonadotropins are known to regulate the seasonal reproductive cycle of fish.

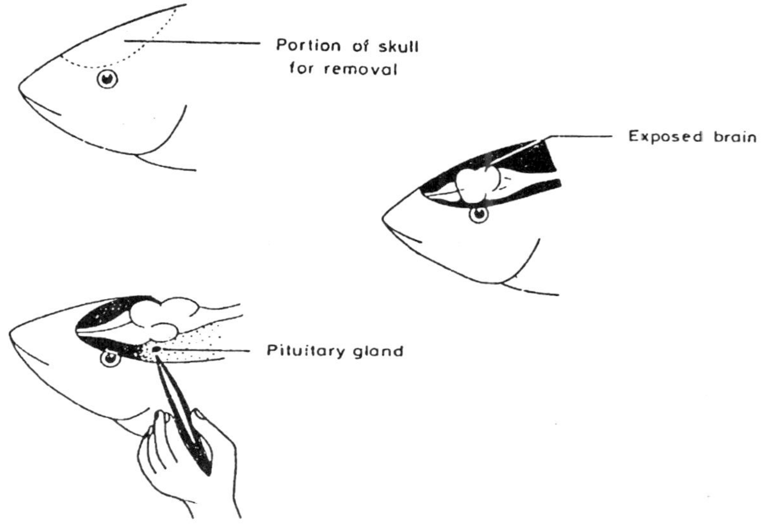

Fig. 6.1 Removal of pituitary gland from fish

PITUITORY GLAND—COLLECTION, PRESERVATION AND HYPOPHYSATION

The pituitary glands are preferably collected from freshly killed matured donor fishes of fish preserved in ice. Generally pituitary glands may be collected from both sexes either belonging to the same species as the receipient fish or closely related species. Generally the pituitary glands are collected from the fish heads in fish markets.

The pituitary glands can be collected from behind the head through foramen magnum after cleaning the brain tissue or brain will be exposed initially and the gland will be collected/removed with fine forceps without any damage. Thus the pituitary glands collected should be preserved immediately in absolute alcohol for dehydration. The alcohol medium is frequently changed in order to keep the glands in good condition. Alcohol is changed for further dehydration and defattening. The glands are then weighed and preserved in fresh alcohol contained in dark coloured phials and sorted either at room temperature or in refrigerator. At the time of injection of breeders, the required quantity of glands are taken out of the phials and the alcohol allowed to evaporate. The glands are then macerated with a tissue homogenizer either in distilled water or 0.3%

saline. The gland suspension is then centrifused and the supernatant fluid drawn into a hypodermic syringe for injection. Glycerine extract is found to retain its potency from 9 to 61 days at room or under lower temperatures. In usual practice, the female alone is injected with a stimulating dose of 2-3 mg/kg body weight followed by a second dose of 5-8 mg/kg weight after a lapse of 6 hrs, if required. Two males per female by weight, are given a single dose, each 2-3 mg/kg at the time of second injection to the female if spawning is not effected. Intramuscular injection of pituitary extract is administered with a hypodermic syringe at caudal peduncle or shoulder region near the base of the dorsal fin. Both the injected males and the female are kept together in a breeding 'Hapa' for spawning. Slight alterations in doses may be made depending on the stage of maturity of the breeders as well as on environmental factors. A breeding hapa is a box-shaped cloth container provided with a cover. All the sides of the hapa are closed excepting one through which spawners can be introduced or taken out. The open corner also can be closed after the introduction of breeders in order to prevent them from escaping. The breeding hapas, made of fine-meshed mosquito net cloth, are of varying size groups. These are:

1. 3.66 × 1.88 × 0.9 m for breeders weighing above 4.36 kg

2. 2.44 × 1.22 × 0.9 m for breeders weighing between 1.63 to 4.36 kg

3. 1.83 × 0.9 × 0.9 m for breeders weighing below 11.63 kg

The breeding hapa is fixed to bamboo poles in a pond, river or watershed. About 13 to 23 cm of hapa should remain above the water surface, while its bottom should not touch the muddy pond bed. Ripe males with milt oozing freely and females with soft bulging rounded abdomen and with reddish vent are selected for breeding purpose. A male Indian major carp can easily be distinguished from the female during the breeding season by the roughness of the dorsal surface of its pectoral fins compared to smooth surface of the female. Healthy breeders, weighing 1.5 to 5.0 kg always preferred. In order to maintain the breeders in the healthy condition, they are often fed with mustard oil cake and rice bran at the rate of 1% of their body weight for a few months prior to breeding season. The stocking rate of the brood fish in ponds is 100 to 2000 kg/ha.

SPAWNING

The carps breed within a wide range of water temperature at

24-35° C belt below 30° is optimum. The spawning takes place 3-6 hrs. after second dose of injection. The eggs are examined to evaluate the percentage of fertilisation. The fertilized eggs are transparent and look like pearls due to swelling up of eggs. Unfertilized eggs become opaque and whitish. The carp eggs are nonfloating and non-adhesive in nature and round in shape. The size of the eggs of carp varies from 4 to 6.5 mm in diameter.

ESTIMATION OF EGG QUANTITY

The eggs are collected from the hapa by means of cup or tray or beaker and transferred to the buckets. The breeders are also removed from the hapa, their weights are noted. The difference in weights reveals approximately the number of eggs laid. The eggs are kept in a rectangular piece of close meshed mosquito net and allow the water to drain off. The eggs are measured in a beaker, mug or cup of known volume and transferred to hatcheries. Thus estimation of total quantity is made from total volume of the eggs measured. Percentage of fertilization can be arrived at by counting the number of fertilised eggs from egg samples of 1 ml measure.

Environmental factors such as light, temperature, water condition, meterological conditions, etc., are known to play important role in stimulating the release of pituitary gonadotropins within the organism and thereby controlling reproduction of fish.

INDUCED BREEDING IN PRAWNS

REPRODUCTION IN WILD PRAWNS

Generally, the male prawns are smaller than females. The female has a characteristic thelycum (sperm stocking organ) which is located on the ventral side of the cephalothorax between the 4th and 5th walking legs. The thelycum is a closed type in *Penaeus indicus* and *P. monodon*. The ovary is found on the dorsal side of the animal along its entire length. In the fully ripe female, the ovary is dark-olive green in colour and has a lateral expansion in the first abdominal segment. The sexual organ is the petasma, which is the modified structure of the first pair of swimming legs attached to the ventral side of the first abdominal segment. The two sperm ducts from the male testis open at the base of the fifth pair of walking legs. The terminal portion of the sperm duct is enlarged to form the sperm pocket is stored. The spermatophore containing the non-motile

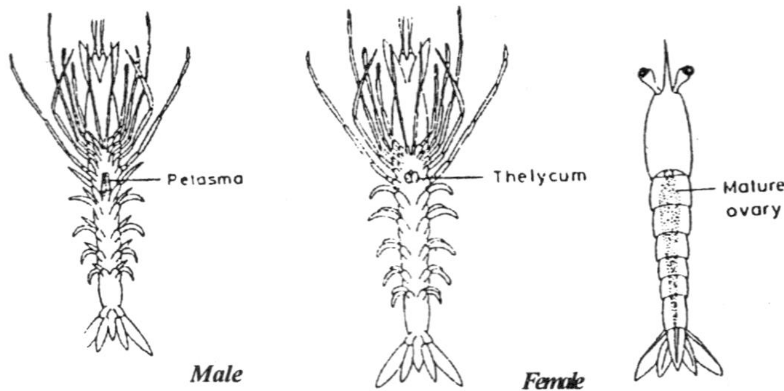

Fig. 6.2 Reproductive organs of Penaeus monodon

sperms is visible as a white mass at the base of the fifth pair of walking legs in mature males.

The mating takes place between a newly molted soft female with immature ovary and a mature male in the intermolt phase and the spermatophore is tucked safely inside the seminal vesicle of the females, the thelycum with the help of petasma. The emptied spermatophore is retained in the thelycum during the intermolt period or until the female prawn molts and then it is discarded along with the moulted cuticle at the time of molting. It is also found that the sperms which are stored in thelycum for some time acquire the capacity to fertilise eggs. Hence direct mixing of sperms of male and eggs stripped from female does not lead to fertilisation of eggs. The technique of removing spermatophore from male by electroejaculation and transfering the sperm pocket to thelycum of a newly molted female developed recently has been found to be very useful in inseminating *P. monodon* artificially. Such prawns have been found to mature and spawn viable eggs in the laboratory. During spawning, the female releases simultaneously eggs from oviduct and sperm from spermatophore and fertilisation takes place in the sea. The fertilised eggs sink to the bottom and come up when the water is agitated.

ROLE OF EYESTALK IN THE REPRODUCTION OF PRAWNS

In prawns, the egg production (vitellogenesis) is mainly influenced by the neurosecretory centres located in the ganglia of eyestalk (X-organ sinus gland complex), brain and thoracic ganglia. While the X- organ sinus gland complex of the eyestalk inhibits vitellogenesis under the influence of its hormone, viz.., gonad inhibiting hormone (GIH), the brain and thoracic ganglia promote vitellogenesis by their hormone, viz., gonad stimulating hormone (GSH). In nature, when the physiological conditionss of the prawns and the environmental factors are favourable, the secretion of GIH is suppressed and thereby, the GSH promotes vitellogenesis. On the basis of this principle, the ovarian development and maturation of gonads are obtained in prawns through the unilateral eye stalk ablation (Removal of one of the eyestalks). It is also worth mentioning here, that removal of both the eyestalks though lead to rapid ovarian growth, spawning does not result under this condition, because the ova are reabsorbed in the ovary, which may be due to the following. The eyestalks apart from GIH, produce other neurosecretory hormones which regulate lipid metabolism and protein synthesis in hepatopancreas; induce hyperglycemia in blood to combat stress; regulate calcium metabolism during cuticle formation; effect water balance during moulting; inhibit production of moulting hormone by Y organ and influence movement of pigments in chromatophores. As these physiological functions are affected by the removal of both the eyestalks, the spawning is also simultaneously impaired. However, the presence of one eyestalk is enough in favouring spawning as it helps in the normal functioning of all metabolic processes. Further it is also interesting to note that the behavioural pattern of the eyestalk ablated females especially their feeding and mating are also not affected.

METHODS OF EYESTALK ABLATION

Several methods are available for the removal of eyestalk. They are:

1. Cutting the eyestalk near the base with a pair of scissors followed by sealing with a pencil type soldering iron (Electrocauterisation)

2. Pinching of eyestalks and squeezing of the eye ball contents cut.

3. Incission of eye ball followed by enucleation of contents.

Among the various methods, electrocauterisation is favoured as it seals the cut instantly, avoiding bleeding and ensures hundred per cent survival.

EYESTALK ABLATION IN PRAWNS

For the procurement of brood stock for inducement, some of the prawns reared in ponds for harvestable size may be allowed to grow for a further period of one year before subjecting them for eyestalk ablation, through the age of *P. indicus* and *P. monodon* at first maturity is only 4-6 months and 5-15 months respectively. The eyestalk ablation is carried out only in female and not in male. For eyestalk ablation, *P. indicus* with a carapace length of 30 mm (Total length; weight >20 g) and *P. monodon* with a carapace length of 50-80 mm (Total length >200 mm and weight >100 g) are normally selected. Before removing the eyestalks, the brood stock are kept in large plastic pools containing filtered seawater of 33 ± 2 ppt. After about 24 hrs, the eyestalks are removed in the females using electrocauterisation technique. The ablated females and unablated males are then kept at 2:1 in larger circular plastic lined pools, in which the seawater is made to circulate through subgravel filter by air lifts. The breeders are fed *ad libitum* using chopped fish or clam meat. The temperature, salinity and pH of the medium are maintained at 28 ± 2° C, 32 ± 2ppt and 8 to 8.2 respectively.

Generally the mating and impregnation are ensured in the prawns if the males and females are kept separately in two tanks and the females are introduced into the male tank when they are about to moult. The separation of sexes seem to enhance the attraction between them facilitating early spawning.

SPAWNING

While the ovarian maturation can be obtained within 4-5 days in the eyastalk ablated *P. indicus* and in *P. monodon* it takes 10-15 days. Subsequently, the spawning starts after 5 and 15 days of ablation invariably takes place during the night. In these species, the number of spawnings in the eye stalk ablated females has been found to be 6-7. *P. monodon* has been found to spawn up to 6 times at intervals of 3-5 days in one intermolt period of about 30 days. There was no marked decline in the number of eggs in the subsequent spawnings though there was decline in the hatching rate in the later spawnings. This decline is mainly due to the reduction in the number and viability of sperms towards the end of the intermoult period. In *P. semisulcatus,* though there is a reduction in the size of eggs at subsequent spawnings, there was no significant difference in hatching and metamorphosis. Further, in this species a shortening of moult cycle and increase in egg number were noticed in ablated females.

The repeated spawnings in the ablated females during the intermoult period (20 - 30 days for *P. monodon*) indicate that one mating and impregnation is enough to fertilise several batches of eggs. However, fresh impregnation would be needed after each moult. In *P. monodon,* two spawnings with a fecundity of 2,77,000 eggs after 3 weeks of abalation and another 2 spawnings with a total of 1,60,000 eggs after 4 days of the second spawning have been recorded. The fertilisation and hatching are, however, at the rate of 25 - 50% and 10 - 30% respectively. In *P. indicus* the fecundity is > 1,00,000 eggs.

Though repeated spawnings occur in the eyestalk ablated *P. indicus* and *P. monodon* after one mating and impregnation, generally, there is no decline in the number of eggs in the subsequent spawnings. However, changes in the quality of eggs have been observed in the eyestalk ablated female in *P. monodon.* This species is found to spawn 4 or 5 types of eggs, viz.,

1. Normal eggs capable of releasing healthy nauplii
2. Eggs which hatch out into abnormal nauplii with deformities
3. Unfertilised eggs with irregular cytoplasmic formations, viz., unfertilised eggs with no segmentation
4. Unfertilised eggs with cytoplasm invaded by bacteria.

Bilateral ablation of eyestalks in *P. monodon* with an interval of about one week between them, has also lead this species to successful spawning in female associated with the release of 6,00,000 eggs.

TRANSPORT OF FISH SEED

Transportation of fish seed from either natural environment or collection spots or hatcheries is considered to be a crucial step in aquaculture, because mishandling of seed causes heavy mortalities. To overcome such kind of mass mortalities, the transport of seed should be properly planned and implemented. Tradtional methods of carrying the seed in earthern pots called 'Hundies' taken either as head loads from the production centre to the market has been an ancient practice in several parts of India, but this type of transport resulted in heavy mortality. At recent times several methods are now available for easy and safe transportation of fish seed.

The reasons or causes attributed for the mortality of live fish at all stages during and immediately after transport are high CO_2 tension and deficiency of O_2 in the transport medium, accumulation of toxic metabolites and other metabolic products like ammonia, strain, heperactivity and exhaustion of fish and also physical injuries, pH of the medium, etc. There are two methods of packing and transporting the fish seed.

1. Open system comprising open carriers, with or without oxygenation/ aeration.

2. Closed system having sealed air-tight carriers with oxygen. For long distance transport with both the systems of transport, spawn and fry are conditioned to make them live in restricted area during transport. The most common method of conditioning is to store fry in a cloth 'Hapa' in ponds.

Generally the fish seeds are very active and need anaesthetisation for transporting them in live conditon for longer distances in good quatities. Otherwise they may exhaust the available oxygen of the transporting media at a shorter time span. So, therefore several anaesthetics were introduced into the market for reducing the metabolic activity of the seed during transportation. But the concentration of various anaesthetics used in differs widely depending on species, size and population density, etc., but they tend to have a common characteristic feature, i.e., considerably reducing the metabolic rate of the seed. This actually leads to the reduced consumption of O_2 during transit and thus facilitates reasonably higher survival rate and withstand long duration of journey in O_2 packed containers in higher stocking densities. All the anaesthetics should have certain features like, it must be a water soluble one, the lethal concentration should be very high, so that, fish do not die accidentally, the dosage required should be low, the fish should tolerate well for several hours at low concentrations, time of induction and recovery should be short.

The modern transportation devices include closed systems like plastic bags, bins, buckets, collapsible plastic pools and fibre tanks. But, most convenient way of transportation of fish seeds is the use of oxygen-packed polythene bags.

Determination of quantity of fish

The quantity of fish seeds for the transportation depends on their size, mode and duration and transport, salinity of the medium and the

TABLE : 6.1. FISH ANAESTHETICS AND THEIR DOSAGES

S. No	Anaesthetics	Dosage	Use	Remarks
1.	MS 222 (Tricaine methane sulphonate)	0.003-0.005 g or ml / 100 ml	Transportation of breeder fish	Expensive ; not ideal for eggs and sperm
2.	Carbonic acid	500 ppm	Transportation of fingerlings	Cheap, safe and easy to use; Immediate
3.	Benzocaine hydrochloride	25 mg/l	Transportation of juveniles	Associated with reduction of Ammonia, CO_2
4.	Sodium amytal	50-170 ppm	Transportation of breeders	
5.	Tertiary amyl alcohol	0.025-0.05 ml/ 100 ml	Transportation of breeders	
6.	Phenoxy ethanol	0.01 - 0.05 ml/ 100 ml	Transportation of breeders	
7.	Methyl pantynol	1 : 10000	Transportation of breeders	Higher dose preferred
8.	Chloral hydrate	0.025 ml/100ml	Transportation of adult and seed	Prolonged induction time; quick recovery
9.	Paraldehyde	0.01 - 0.05 ml/ 100 ml	Transportation of adult and seed	Low induction and recovery
10.	Tertiary butyl alcohol	0.3-0.35 ml/100ml	Transportation of adult and seed	Rarely used

ambient temperature. A simple formula to find out the number of fish for a container is given below:

$$N = \frac{(DO-2) \times V}{C \times h} \quad \text{where}$$

DO = Dissolved oxygen in ambient water in mg/l

V = Volume of water in litres

C = Rate of oxygen consumption by the individual fish (mg/h)

h = Duration of transportation (h).

General guidelines for the transportation of seed

In order to avoid loss of time and money, the transportation of seed should be well planned. The primary need of the containers is ample oxygen for the entire transportation period. The following general guidelines may be adopted for easy and safe transportation of fish.

1. Transportation of seeds must be done during cool hours especially during night or early morning.

2. In order to avoid the rapid warming of water containers before transportation they should always be kept in the shade.

3. Before packing and transportation, the fish seed should be treated with about 1% potassium permanganate solution or common salt solution which serves as a prophylactic measure.

4. After the culture site is reached and before stocking the seed in ponds, they are to be acclimatized with pond water. For this a portion of water from the container has to be removed and replaced with pond water. Subsequently, the entire water is removed from the container and pond water is replaced. The seeds are allowed to get acclimatized for few minutes before releasing them into the culture ponds.

5. All the containers and accessories used in handling and transportation should be very clean.

6. Generally the fingerlings intended for transportation should so as to avoid the damage to the scales and fins.

7

POND CONSTRUCTION AND MANAGEMENT

FARM SITE

An obvious prerequisite for fish culture is the availability of different types of ponds for rearing various stages of fish. For the production of fish of different sizes or required sizes, designing and construction of specific ponds is a basic principle. The primary consideration in constructing a fish farm is the choice of a site which would provide good quality of water retensive soil base and pond sites, having an assured adequate water supply to fill in ponds at any time of the year.

REQUIREMENTS OF A NORMAL FISH FARM SITE UNDER INDIAN CONDITIONS

Particulars	*Requirements of normal fish farm site*
1. Nature of terrain	Non rocky with at least 2 m deep soil
2. Slope of the terrain	Land should be level or gently sloping
3. Physical quality of the soil	Soil fraction should be about 90% of the whole soil, stone and gravel not exceeding 10%.
4. Chemical quality of the soil	pH near neutral. Total nitrogen > 0.1% Total phosphorous > 0.1% Organic carbon > 1.0% Free $CaCO_3$ > 5%

{Cont.}.........

5.	Rate of fall in water level in ponds	Should be less than 1 m per annum
6.	Water table	Should not be far below the pond bottom when the soil is not water retentive
7.	Water supply	There should be a source of perennial water supply nearby, sufficient to meet the requirement of the farm
8.	Biological productivity	Average plankton production per m^3 should range between 10 and 20 ml.

In piscicultural programmes, fish culturist requires different types of ponds rearing various sizes of fishes. Different types of ponds to be digged for rearing fishes. For carp culture, at least three different types of ponds, i.e., nursery, rearing and stocking ponds are required. All these different kinds of ponds differ from each other with respect to its size, area and depth of water. The nursery ponds are small and shallow having an area of 0.02 to 0.08 ha (20 m × 10 m to 40 m × 20 m in size) with a depth of 1.2 to 1.5 m. These ponds are used for raising fry from spawn (5.6 to 25-30 mm). The nursery ponds should preferably be seasonal in nature.

The rearing ponds are slightly larger in size than the nursery ponds. The pond size may vary from 0.08 to 0.10 ha (40 × 20 m to 50 × 20 m) with a depth of 1.5 to 2.0 m. These ponds are used for rearing fingerlings from fry with a size ranges from 25-30 to 100 mm and above. The rearing type of ponds are still larger than the rearing ponds. They differ in size from 0.2 to 2.0 ha with a depth of 2.0 to 2.5 m. It is desirable that the stock pond is more than 0.5 ha in area. These ponds are utilised for stocking fingerlings in order to obtain table size fish. They are called stocking ponds and are perennial.

All the ponds in general should be rectangular in shape. The pond width should not exceed 40 m. This will facilitate longer run for fish to achieve healthy growth and also smaller nets and lesser number of people are required to catch the fish. Larger ponds are difficult to manage in several aspects including the controlling of aquatic weeds.

The number of likely factors to be taken into consideration for farm pond construction can be broadly divided into 3 catagories.

1. Site selection
2. Designing and
3. Construction.

SITE SELECTION

The economics of fish pond construction and the productivity of pond largely depend on the selection of suitable site for culture activity. There we have to take care in selecting the best possible site for construction of culture pond. The more important aspects of fish farm construction are topography, soil type and water supply.

Topography

Topography or layout of land is the surface feature of water shed. The ideal topography of a fish site is gently sloping terrain of a wide valley or basin surrounded on three sides with high lands and with a narrow outlet on the fourth, as it would keep the cost towards excavation of early to the minimum. Such a place can be chosen for constructing the farm, provided the desirable type of soil and suitable water supply are available. Such an area can be easily converted into a large pond by erecting an embankment for closing the outlet. The type of farm to be constructed in a terrain would depend upon its topography.

Soil type

The soil should be free from vertical and lateral seepage. Such impervious soils are ordinarily heavy clay, silty clay, clay loam, etc. Sites to be avoided are : rocky out crops, limestone areas. Barren, slightly sandy, marshy or water logged areas should be chosen as suitable sites as such areas can be put to no other profitable use. Pond construction on predominantly sandy soils would, however, be extremely expensive and involve operational cost and should therefore be avoided. Broadly soils can be classified into 3 catagories, i.e., sand, silt and clay, depending upon the size of the grain of the soil. Sands drain readily while silt and clay are difficult to drain. Clayey soil has a very good capacity of retaining water and in this respect such soil is the best of all. Clay can be identified by its sticky and plastic properties and special odours. When dry, clay forms hard lumps which cannot be broken down and powdered between fingers. Clay is very adsorbent and it tends to hold free water. On the contrary, a sandy soil if squeezed in hand when dry, will fall apart when the pressure is released, while moist, it will form a cast but will crumble when touched. Ponds should not be constructed over sandy soils because sand is too porous to retain water. If the selected site is poor in water retentivity, clay has to be brought from elsewhere and a c' / coating and puddle core has to be provided to make the ponds retain water. Large quantities of water and cowdung help in improving the water retention capacity of poor soils.

Water supply

The third requisite for selecting a site is the availability of an adequate and dependable source of water. The usual sources of water supply are : reservoirs, streams, springs, canals, surface runoff and wells. A source of water supply to fill the ponds against losses due to seapage and evaporation is therefore essential and should be ready at hand. Areas with sufficient rainfall, high water tables and water retentive soils do not need a stand by source of water supply.

DESIGNING

Construction of Fish Ponds

While constructing any pond viable for fish culture, one must see that the construction phase is of a short period and the production phase is of longer one. Any suitable area can be converted into an earthern pond by digging 0.5 to 1 m deep. Rectangular ponds are suitable than square or circular ponds for easy netting operations.

A typical earthen pond will have the following structures, viz., bunds or dikes, harvesting pit, inlet, outlet and core trench.

Bunds

Bundhs are the protecting structures of fish ponds and are essential for both drainable and undrainable fish ponds. They may be of three types, viz., (a) main bund or peripheral bund which is essential for larger fish farms enclosing a number of fish ponds of varied sizes., (b) bunds holding water on one side and, (c) bunds dividing two adjacent ponds. The main bund is the highest part and is constructed in the deepest part of the pond area. The bund holding water only on one side is also constructed as main bund, but it is not constructed on the deepest area of the pond.

SLOPE

The life span and strength of any bund depend not only on the quality of soil but also on its slope and crown width or crest. The slope may be defined as the ratio of horizontal increase in the base of the bund from the point of perpendicular to the top edge of the bund to the edge of the base of the bund on the same side per unit length. For example, if the height of the bund is 1 m and basal width of its one side is 1.5 m, then the slope is said to be 1:1.5. The slope of a bund is based on the quality of soil used for its construction and required height. Normally, good clayey soil makes steeper slope than the sandy-clayey or sandy soil. For ordinary earthern ponds of less than 0.5 ha, the wet side slope may be 1:1.5 and the dry side slope 1:1. The crown width also depends on the height of the bund. However, a minimum of 1 m crown width is invariably needed for any bund. For a bund of 2 m height, it may be around 2 m. If the main bund of a fish farm is to be motorable one, a minimum crown width of 3 m is essential.

Berm

If the area of the pond is more than 0.5 ha, a platform-like space between bund and watery area known as berm or bench line should be made available. The width of the berm may vary between 0.5 to 1.0 m, depending on the height of the bund and the size of the pond. The berm apart from serving as a walkable space for fisherman also protects the bund from direct contact with water.

CONSTRUCTION OF POND

It will always be economical if the bund is formed out of the earth removed from the enclosing area. Hence, before constructing any bund, it is better to calculate the total quantity of the earth required for constructing bunds for all the four sides of a fish pond and the area to be dug in the enclosed area for getting the earth.

Determination of quantity of earth for the construction of bund

To find out the total quantity of earth required for the construction of bunds, it is necessary to determine the quantity of earth required for one metre long bund with the proposed specifications regarding slopes, height, crown width, etc. The following may be taken as example:

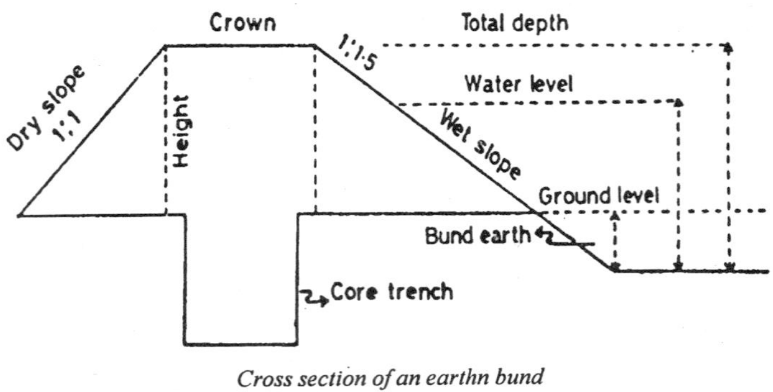

Cross section of an earthn bund

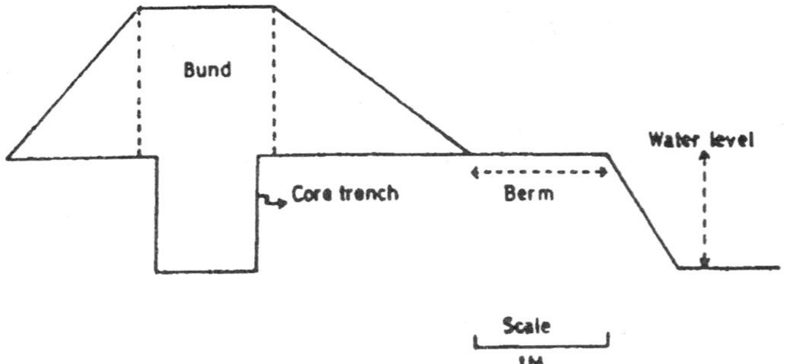

Cross section of an earthen bund showing berm

Fig. 7.1 Bund and Berm

Area of pond	=	1,000 m^2 (50 × 20 m)
Crown width	=	1 m
Dry side slope	=	1:1
Wet side slope	=	1:1.5
Bund height	=	1 m

Earth required for one metre long bund =

$$= \frac{1 \times 1 \times 1.5}{2} + (1 \times 1 \times 1) + \frac{1 \times 1 \times 1}{2} = 2.25 \text{ m}^2$$

(The clay required for core trench is not calculated in this calculation)

Total length of actual four bunds (20+50+20+50) = 140 m. Total earth required for 140 m long bund = 140 × 2.25 M^3 = 315 m^3. In order to form the four bunds of the above specifications, it is enough, if earth is taken from a depth of 0.315 m from the enclosed area, i.e., 1,000 m^2. However, this is only an estimate and invariably 10 per cent or more earth may be needed in order to meet the earth requirements of corner of the bunds and compensate the loss due to compaction, particularly after the rains. Considering the above, a depth of 0.35 m may be dug and earth be removed.

BUND FORMATION

Sandy or gravelly soil is not suitable for bund construction. Further, a soil with high clay content may also form fractures in the bunds. Hence, a mixture of silt, sand and clay in the ratio of 1:3:2 should be used in the construction of firm bunds. After all the area data relating to the bund are gathered, a design is made for the pond construction. Using the design, the bund lines are pegged along with the contour of the pit from which the required earth is taken out.

The axis of the bund with a base of known length is first pegged and their sides are demarcated. After pegging the bund line, the debris, roots, stones, etc., are removed from the marked area. Subsequently, the top fertile soil of 5-10 cm depth is also removed. This may be heaped in a nearby area for future use. If the soil of the bund area is of pervious material, then the formation of a core trench is a must. The core trench depends on the base width and height of the bund. For example, a bund of 3.5 m base width may have a core trench of 0.75 m width and 0.75 m depth. After the area pertaining to the core trench is marked, it is dug and the clay brought from outside is packed in layers and compacted. The core trench is mainly to prevent seepage of water through bund and to strengthen it, and it may be avoided if the pond is not deeper than 1-1.5 m and soil is clayey and impervious. The profiles of the bunds are then made with long and thin wooden pieces, which may reflect the height and crest of the bund. The gradient of the slope is indicated by strings stretched between the top of the wooden pieces and peg, marking the side of the wall base. The final phase of the bund construction is fitting out, where the bits of earth are properly packed according to the bund profile setup. Subsequently, the top soil heaped apart is spread over the slope which would not only enhance the grass production but would also avoid the erosion of the bund. To hasten the growth of grass, adequate amount of turf may also be placed on the slope. Finally, the pit should be levelled. Before filling the pond with water, its bottom should also be levelled and cleared of debris, stones, etc. A gradual but insignificant slope from one end to the other end in the case of undrainable fish pond may be provided at the bottom. This will help, particularly during summer or harvesting period, when water level goes down considerably and the harvestable fish can find refuge in the deepest area with water.

INLET AND OUTLET

These are required especially for larger ponds, in order to ensure a smooth water supply and drainage. Further, they are helpful in preventing the entry of wild fish from outside and escape of the fish from inside the pond. The size and shape of these structures largely depend on the area and waterspread of the ponds. For a pond of 0.1-0.2 ha area, an inlet or outlet pipe of about 15 cm diameter will be ideal, as a pipe of less than 10 cm may lead to clogging. The outlets or drains are invariably installed at the lowest plane of the pond. However, for larger ponds (1-2 ha) the pipe or culvert should be of around 25 cm diameter. The inlet/outlet pipes are to be provided with suitable screens or other devices, so as to keep off wild fish as well as prevent escape of cultivated fish. However, to stop the flow of water through the inlet and outlet pipes, in small and medium sized fish ponds, plugs of wooden slumps or gunny cloth also be used as and when needed. For larger fish ponds the construction of concrete inlets and outlet are highly useful.

While designing the fish farm with nursery, rearing and stocking ponds the following points should be borne in mind in addition to the survey reports and maps.

1. The ratio of water area among nursery, rearing and stocking ponds largely depend upon the purpose of the farm. In case of fish farm is meant for fish seed production only, nursery and rearing ponds can be constructed leaving a nominal area for stocking the breeders. If the farm is aimed to produce the table size fish then, only the stocking ponds needs to be constructed.

2. Nursery, rearing and stocking ponds should be designed and constructed according to the sizes and depths. All the ponds should be designed in such a way that they are rectangular in shape and the width of the pond should not exceed 40 m.

3. The slope of the dyke depends on the soil type and its height.

Generally, the slope of the dyke of a nursery, rearing and stocking pond may be 1:1, 1:1.5-2.0 and 1:2.0-2.5, respectively.

4. Embankment of the fish pond should be designed in such a way that its height from the bottom of the pond should be 2.7 to 3.7 m. A 0.6 m free board should be left between the water surface and crest of the embankment. The free board is essential to prevent the pond spilling over due to sudden and exceptional downpours.

5. Ponds should have controlled inlets and outlets, so that these can be easily filled up and also drained.

6. The layout of the ponds should be as per the land configuration and the contours. The deeper ponds should be placed on the lower contours, so that lesser earth work is involved.

After designing the fish farm the following points .. considered at the time of construction of a fish farm:

1. From the point of view of pond management the time of construction of fish ponds is an important factor. The most desirable time to complete the construction of fish farm is in late summer, so that the pond will be immediately filled with rain water during the monsoon season.

2. All trees and shrubs and other vegatation should be cleared, or else after decay these constitute dangerous pockets where leakages develop. The top soil may be scrapped off and collected at one place to be used later for dressing.

3. The dyke bases are marked off and if the soil is impervious, excavation may be started to fill the area.

4. A firm and sufficiently wide berm adds to the stability of the embankment. If the berm is not wide enough the side slope of the pond will give way and collapse due to the vertical load of the embankments. A wide berm is always heplful for operating the nets in the ponds. For an ordinary pond the berm width can be kept at 1.2 m.

5. Proper drainage arrangement should be provided so that the fish pond can be emptied and filled whenever necessary.

POND MANAGEMENT

Scientific management techniques of farm ponds are essential to attain maximum yield. Freshwater carp culture in India is a three stage culture system, which includes the management of nursery, rearing and stocking ponds. Management practices evolved can be broadly grouped under pre-stocking, stocking and post-stocking operations. Pre-stocking management practices include the maintenance of ponds, removal of aquatic weeds, eradication of undesirable and predatory organisms, improvement of pond water and soil quality and fertilisation. Stocking phase consists evaluation of carrying capacity of the pond, relative density and combination of selected species.

Post stocking management practices mainly comprise the supplementary feeding, fish health monitoring and harvesting pattern of the crop.

NURSERY POND MANAGEMENT

The nursery pond is mainly for growing spawn and hatchlings to advanced fry of the size of 2.0 to 2.5 cm for a period of about 15 days. The management of nursery pond includes the clearing of algal blooms and weeds, eradication of unwanted animals by using appropriate poisons, liming, fertilising, releasing spawn and their nursing, supplying the artificial feeds and finally harvesting.

Algal Blooms

Algal blooms are invariably caused by unicellular and filamentous algae and may impart any to the water, like green, yellowish-brown, bluish-brown, etc. The groups of algae which are responsible for the development of blooms include Chlorophyceae (Green algae), Xanthophyceae, Bacillariophyceae, Chrysophyceae, Cryptophyceae, Dinophyceae, Euglenophyceae and Myxophyceae. While the blue-green algae are responsible for the development of persistent blooms, others cause temporary blooms.

Control of algal blooms

1 ppm Copper sulphate solution or Copper sulphate at 800 g/ha packed in cloth bags and immersed in ponds is known to eradicate algal blooms, particularly of blue-green algae. It is further recommended that

it is necessary to apply this chemical once a week for a number of weeks if needed. Further 125 ppm sulphuric acid or 0.5 ppm Simazine may also control the Microcystis bloom very effectively.

The algal blooms are very effectively controlled by introducing the duckweed, Lemma which forms a shade over the algae and prevent their growth. Silver carps, the algae feeding fish may be introduced for the eradication of fish.

Aquatic weed management

The unwanted aquatic weeds have several disadvantages, viz.

(a) limiting the space for culturable fish
(b) causing lot of disturbances in the oxygen concentration
(c) hampering the netting operations
(d) causing siltation
(f) harbouring unwanted predatory fishes and molluscs.

So far 5 types of weeds are recorded

(a) Algal weeds : These are the filamentous algae form a mat on the surface of water. Ex. Spirogyra, Nitella and Chara.

(b) Floating weeds: The leaves of these weeds float above the surface and roots underneath. Ex. Eichornia, Lemma, Pistia, Azolla.

(c) Emerged weeds: The roots of these plants are attached to the bottom and leaves freely float on the surface. Ex. Nymphaea, Typha.

(d) Submerged weeds: These plants are rooted or rootless in the water column. Ex. Utricularia, Ceretophyllum, Vallisnaria, Hydrilla.

(e) Marginal weeds: These are mostly rooted and are seen either on the marginal areas of the land or on the adjoining land. Ex. Marsilia.

Control of Aquatic weeds

Mechanical methods such as cutting, dragging, dredging, raking, etc., may be employed for the collection and disposal of the aquatic weeds.

Chemical methods for the control of aquatic weeds are as follows:

Blue green algae and algal weeds : 1 ppm Copper sulphate
0.5-1 ppm Simazine
0.1-0.3 ppm Diuron (1 kg/ha)
0.2 ppm Paraquat (1 kg/ha)

Floating weeds : 80% 2-4 Dichlorophenoxy acetic acid (5-7 kg/ha)
Simazine (5 kg/ha)
15 ppm Ammonia
0.2 ppm Paraquat (1 kg/ha)

Emerged weeds: 5% Dalapon (2-2 dichloropropionic acid) (5-10 kg ha)

Submerged weeds: 15 ppm Ammonia
3-5 ppm Simazine
4 ppm Sodium arsenite
500 ppm Superphosphate

Marginal weeds : 0.2 ppm Dequat (1 kg/ha)
0.2 ppm Paraquat (1 kg/ha)

Biological methods including the introduction of Grass carp and Common carp for the eradication of unwanted aquatic weeds.

CONTROL OF PREDATORY AQUATIC INSECTS

Different types of insects either as larvae or adults may not only prey on fish hatchlings but also compete for food. The most important insects posing problems to the fish ponds include the water strider Limnometra, backswimmer Notonecta and Anisops, water scorpions, Laccotrephes, diving beetles, Cybister, Hydaticus, scavenging beetles, Hydrophilus, etc.

Control methods including repeated dragging through water of a fine meshed cloth net may help in the removal of large numbers of these insects to a certain extent. The application of vegetable oils thereby causing the respiratory failure in insects is also another measure of controlling. Different chemicals are also used to eradicate these insects include BHC (0.1 ppm), Endrin (0.001 ppm), Dieldrin (0.5 ppm) and Quicklime (204 ppm).

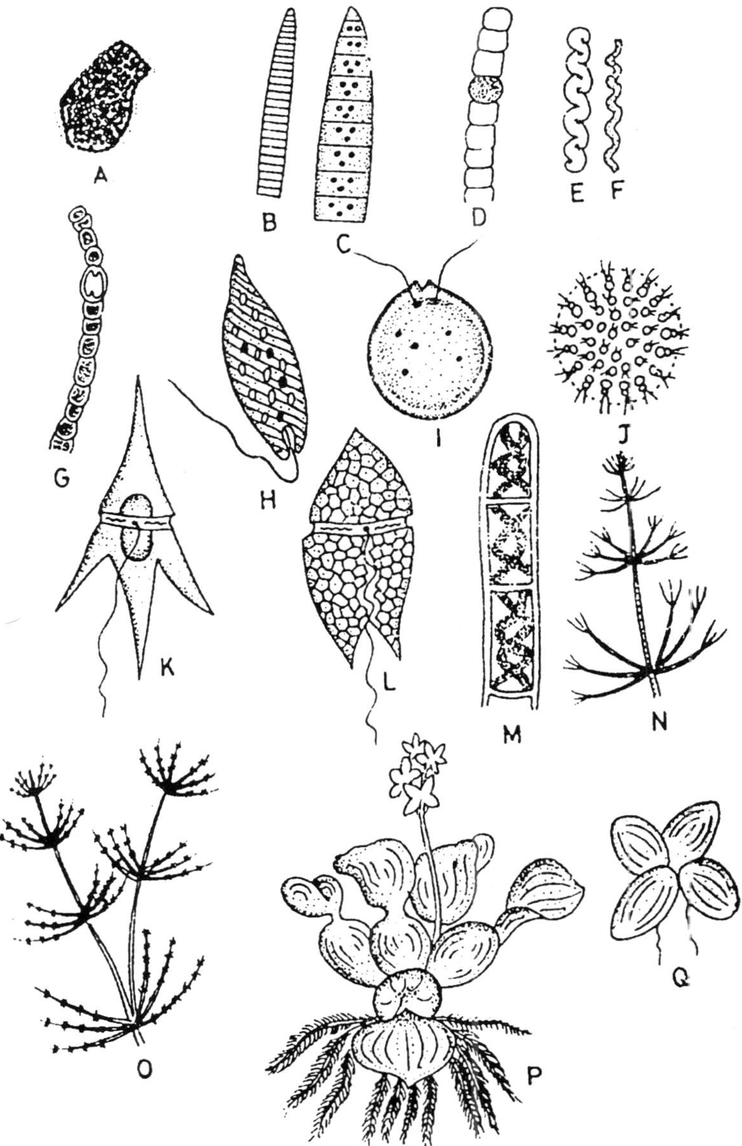

Fig. 7.2 Species of algae causing blooms and aquatic weeds

A. Microcystis, B. & C. Oscillatoria, D. Anabaena, E. & F. Spirulina, G. Nostoe, H. Euglena, I. Chlamydomonas, J. Volvox, K. Ceratium, L. Peridinium, M. Spirogyra, N. Nitella, O. Chara, P. Eichhornia, Q. Lemna.

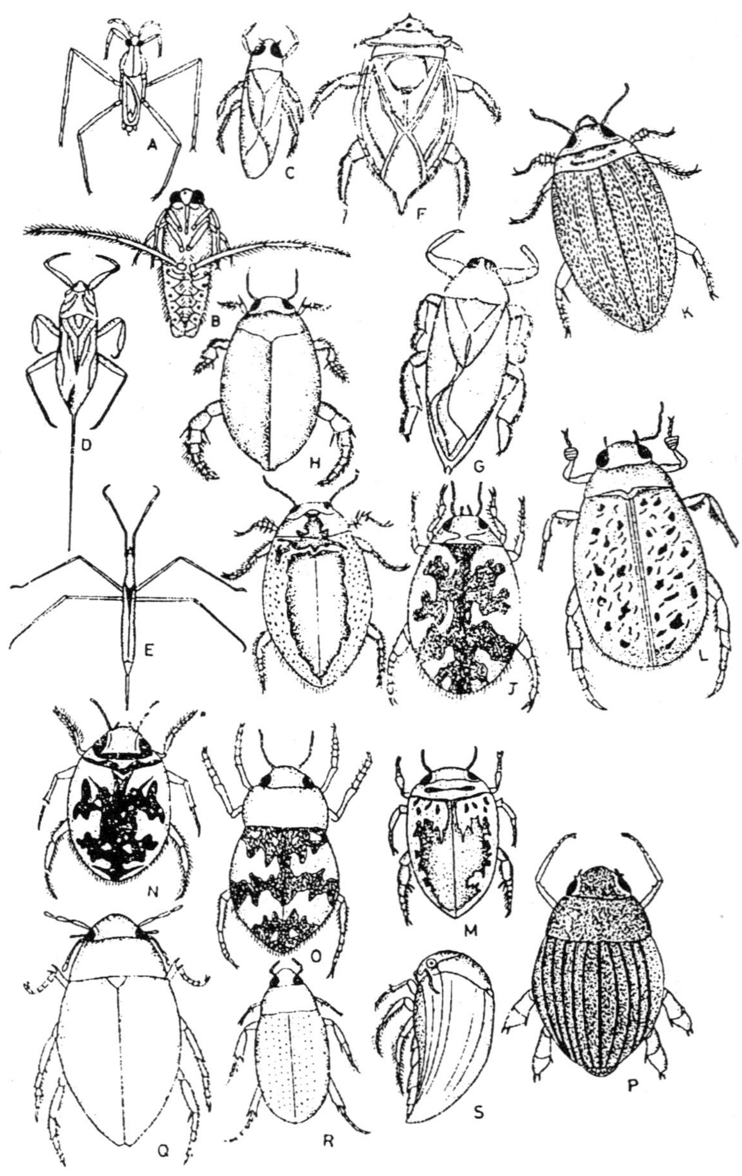

Fig. 7.3 Predatory insects

A. Limnometra, B. Notonecta, C. Anisops; D. Laccotrephes; E. Ranatra; F. Diplonychus, G. Lithocerus, H. Cybister, I. Hydaticus, J. Sandracottus, K. Rhantaticus, L. Eretes, M. Laccophilus, N. Hyphoporus, O. Peschatius, P. Dineutes, Q. Hydrophilus, R. Sternolophus S. Regimbartia.

Control of other predators

To control frogs and their tadpoles, snakes and birds, etc., suitable measures have to be followed. By applying quicklime, the eggs of frogs can be killed. Birds like herons and crans are invariably controlled by shooting, scaring or netting.

LIMING AND FERTILISATION

Proportianate quantity of organic manure in the form of dry cowdung for nursery pond is to be applied. Overdoses may have adverse effects on the growth of fish fry, particularly in causing the gill rot and oxygen depletion. Before application, the organic manure is made into a paste with water and applied at different parts of the pond. To ascertain uniform distribution of organic manure, the pond is dragged using the net. When the organic manure gets dispersed or finely distributed in the water column, the reducing bacteria present in it decompose the organic matter with the help of oxygen and convert it into potassium, nitrogen, phosphorus and carbondioxide for the use of plankton. The application of commercial lime at the rate of 300 kg/ha in combination with organic manure is known to improve the quality and fertility of the water in the nursery.

The stocking rate of nursery depends on the pond fertility, depth and local climatic conditions, etc. A stocking density of around 1 million/ha is generally recommended.

Supplementary feeding is also essential for the growing spawn. A mixture of powdered groundnut oilcake and rice bran in the ratio of 1:1 is preferred.

REARING POND

All the rearing ponds should be located near the nursery so that the transportation of fry will be easy. Like nursery pond, the rearing ponds also seasonal, where fry are grown to fingerlings of 10-15 cm size for a period of about 3 months. Most of the management practices are similar to that of nursery pond.

FERTILISATION

The rearing pond is first fertilised with organic manure (cowdung) at the rate of 10,000 - 15,000 kg/ha. This would enhance the production of fish food sources, viz., the phyto - phyto and zooplankton which would inturn, promote the growth of the fry. As there may occur depletion of plankton in two to three days after stocking the chemical fertlisers, viz., urea and superphosphate may be applied at the rate of 40-80 kg/ha at every 15-20 days interval.

STOCKING

The stocking density of fry in the rearing pond may be between 5,000 to 1,50,000 /ha in the following ratios.

Catla, rohu and mrigal 1:2:2

Catla, rohu, mrigal and common carp 3:4:1:2

Catla, rohu, mrigal, silver carp, grass carp and common carp 2:4:4:4:3:3

FEEDING

When the production of fish food organisms, viz., plankton goes down in the rearing ponds, addition of supplementary feed to the growing fish is essential to enhance the growth of the different species. Normally the powdered groundnut oil cake and rice bran may be mixed in the ratio of 1:1 and soaked in water for some time. Then it is made into small balls for the purpose of feeding.

PRODUCTION POND

The ideal size for the production pond for profitable fish culture is 0.1-2.0 ha. Here, the fingerlings are stocked and grown for about one year or until they attain a marketable size. The production pond for fish culture is to be fertilised both with organic and chemical fertilisers. The organic manure is of three types :

(a) liquid manure (urine and sewage rich nitrogenous matter)

(b) farm manure like cowdung, pig dung and poultry manure) and

(c) plant manure such as green manure, compost, mahua oilcake, etc.

Depending on the soil characteristics such as pH, organic carbon content, the quantity of organic manure should be determined.

STOCKING

Fingerlings of the size of 4-7 cm are preferred for stocking the production ponds after 10-15 days of initial fertilisation. Prior to the stocking, the fingerlings are to be acclimatized to the water of the production pond. The fingerlings are given a dip treatment with 2% potassium permanganate solution, in order to safegaurd them from parasitic infections. Depending on the fertility of the water the stocking density also varies. The stocking density also ranges from 2,000 to 10,000 /ha.

8

CULTURE SYSTEMS

With the rich resources in the country, there is a tremendous scope for the development of aquaculture. Vast resources in the country have also suited for the enhancement of culture practices. Based on the type of culture, the culture systems can broadly be classified as Freshwater culture, Brackish water culture and Mariculture.

FRESHWATER AQUACULTURE SYSTEMS

The freshwater pond culture has a long history in India and is traditionally confined to some states. In the traditional system, ponds are usually stocked with indigenous major carps like *Catla*, *Rohu* and *Mrigal* and exotic carps such as Grass carp, Silver carp and Common carp. Usually. they are stocked in spawn, fry or early fingerling stage at high densities. The composite fish culture system practiced in India is almost synonym of multispecies fish culture followed in China. In India, both indigenous major carps and exotic carps are stocked in various to obtain high production per unit of water body. Fast growing compatable species of fish of different feeding habits, or different weight classes of the same species, are stocked together in the same pond so that all its ecological niches are occupied by the fishes. In fact, there does not exist any competition between different species but each species may have a beneficial influence on growth and production of the other. For instance, the exotic grass carp by consuming aquatic vegetation converts plant matter into fish flesh and its excreta fertilizes the pond which benefits all other species. Composite fish culture technology involves management of culture pond, water quality management, control of aquatic weeds, eradication of unwanted fishes, fertilisation on the basis of nutrient composition and supplementary feeding, and harvesting of fish at a suitable time depending on the growth performance of each species. In the growout, either in monoculture or in polyculture, the basic objective is the production of optimum quantity of the fish with a minimum cost. However, there are a number of interdependent factors that affect the productivity and cost. Stocking rate or the density of fish in the pond, the quality and quantity of food produced by fertilization or artificial feeds used in the pond, water temperature, availability of oxygen and accumulation of metabolites in the pond are some of the factors that influence the growth rate and production. The size of the fish at the time of stocking and the duration of the culture and the size at the time of harvesting are the otherfactors which influence the growth rate and production, respectively. The most common means of harvesting the freshwater fish culture ponds is by seining the pond, however the ponds may be with harvesting sumps. It is always preferable to harvest by draining, especially, when the pond contains common carp and silver carp.

Freshwater Prawn culture

Among the increasing number of aquatic animals that have been coming *Macrobrachium rosenbergii* is considered to as the suitable prawn for freshwater farming in ponds. It occurs commonly in most of the rivers, lakes, irrigation canals as well as estuaries. It is omnivorous and diet includes insects, algae, grain, seeds, small moluscs and crustaceans, flesh of fish and other animals. It readily accepts, supplementary feeds consisting of plants and animal products. It is a fast growing species and attains maximum length of 300-320 mm. Male prawns are considerably longer than the females. Spawning of *M. rosenbergii* takes place in the gradient zones of estuaries. The young post-larvae ascend the rivers. Migrating juveniles are caught and reared in ponds. *M. rosenbergii* female prawns are reported to lay 60,000 to 1,00,000 eggs during one spawning. Eggs are slightly elliptical, bright orange in colour and before hatching they become grayblack. Eggs are incubated in the brood pouch and incubation time varies between 15-24 days. Egg hatching process is completed normally in one or two nights. The freshwater prawn larvae also known as zoeae are planktonic and swim actively tail first, ventral side upper most. They require brackishwater for their survival. There are eleven microscopically distinct stages during the larval life of the giant freshwater prawn. The first stage is just under 2 mm long and the 11th stage it grows to over 7 mm long. Larval rearing is the most critical phase in the freshwater prawn hatchery system. Monoculture of this prawn at varying stocking densities resulted in the production of 600 kg/ha in 4 months growout period.

Fish culture in Paddy fields

In India, rearing of fish in paddy fields is a traditional practice. Integrated fish and paddy cultivation helps in resource utilisation. When the rice fields are inundated and retain water for 3-8 months a year can provide additional crop of fish to the farmer, without decreasing paddy production. Essential requirements of a paddy field suitable for fish culture are the construction of channels or trenches inside the field. Major Indian carps Catla, Rohu, Mrigal and exotic carps are suitable for culture in rice fields. Freshwater prawn *Macrobrachium rosenbergii* and fish *Puntius* sps. could be cultured along with paddy. Fish culture in rice fields is conducted in four types, viz., catch on irrigated rice fields, secondary crop method, in the interval method and continuous method.

Brackish Water Culture

Vast area with rich resources suitable for brackish water fish farming is available in our country. The coastal area of Bay of Bengal and the Arabean Sea are very much suitable for development of brackish water aquaculture. The bheri fish culture in Sunderbans of West Bengal, the prawn culture in Pokkali paddy fields of Kerala and the culture in coastal lagoons are the most important types of brackish water fish culture practised in India. The different organisms which are cultured in brackish waters are mostly synonomous with the organisms which are cultured in the marine water. Usually, both mariculture and brackish water culture are called Coastal aquaculture.

Mariculture

Systematic and harmonius blending of Sea by farming with traditional or modern capture methods of marine fisheries is called Mariculture. Otherwords, the term mariculture means organised culture of marine organisms in sea water, i.e., rearing of marine forms in the marine environment itself. In general, the edges of the oceans and inshore bays are selected for mariculture activities. The first attempt was made at Mandapam camp by CMFRI in 1958-'59 with *Chanos* culture.

Several plants and animals are being cultivated in marine waters. Different plants and animals which are taken up for mariculture are classified as sedentary and free-swimming organisms. Various organisms such as oysters, mussels and clams are the best examples for sedentary organisms and the fin-fish and prawns are the important free swimming organisms.

Oyster culture

Culture of molluscs, especially bivalve molluscs, is unique from several points of view. Being sessile and low trophic level filter-feeders for most of their lives, molluscs can be raised at relatively low cost. Culture is carried out mainly in open waters, often in the natural habitats with young ones (spat) collected from the very same environment. Several culture methods are prevailed for culture of oysters like bottom culture and off-bottom culture methods which include stake/stick culture, stone bridge culture, typical rack culture and suspended culture systems such as strings, ropes or trays and long lines. *Crosstrea rivularis* and *Ostrea edulis* are the important ones in the edible oysters and *Pinctada fucata* is the common pearl oyster belonging to genus *Pinctada* sps. which are the prominent ones available in the Indian coasts. Fine pearls are produced from the peal oysters and the culture activity of these organisms is paid much attention.

Crab culture

High consumer preference and market price have led the aquaculturists to develop considerable attention to culture crabs though the culture techniques are not still much developed. As a number of species of marine crabs are edible, the biggest and the most highly esteemed species suitable for coastal aquaculture is *Scylla serrata*, commonly known as mud crab, green crab or mangroove crab. The most important disadvantage in crab culture is pronounced canninbalism during the larval and adult stages. In most of the coastal fish ponds, the culture of serrated crab forms a subsidiary crop. The young ones are collected with the tidal water and are grown for a period of about six months till they reach marketable size. For polyculture in brackish-water ponds, crabs are used with milk fish or shrimp, but bamboo or plastic fence has to be erected on the dikes to prevent escape of the crabs.

Lobster culture

Larger consumer preference and market demand led the mariculturists to divert their attention to the culture of lobsters. For a number of years the attempts on the possibility of lobster culture were not that much successful. Attempts were made in rearing the post-larval stages of the spiny lobsters, *Panulirus* sps. and *Jasus* sps. The protracted larval development, nature of food required by the different stages of the Phyllosoma larvae, the long time the juveniles take for growing to a marketable size and pronounced canninbalism have resulted the culture uneconomical for commercial culture.

Mussel culture

Mussels are traditionally a well accepted aquatic food in many countries. It is only in the last few years, the potential for their culture has been identified and received significant attension. The most important one in the point of view of aquaculture is the blue mussel, *Mytilus edulis,* which is widely distributed and adapts itself to a variety of ecological conditions. The muscles are also cultured in several ways, namely, bottom culture, bouchet culture, raft culture and offshore culture methods.

Culture of eels

Eels are of delicacy in our country, whereas in western Europe and Japan it is having high demand. Traditionally, the eels are cultured in stew or holding ponds. Though the artificial propogation of eels has made some progress, the culture of eels still depends on the natural seed collection (elvers). The important eels identified for culture practices belong to *Anguilla* sps. These are known to be catadromous species and migrate from rivers and other inland *water bodies into the sea for breeding. The most common method* of eel culture is the pond farming method. The ponds being comparatively small, may be of still-water type with frequent exchange of water or with running water facility. The culture of eels in net enclosures in sheltered bays, back waters, etc., were also tried to reduce the capital costs of pond construction and to avoid the problem of water supply. A very high production rates of 26 tons/ha have been reported in recent years with intensive farming systems, particularly using heated water and running water systems.

Culture of aquatic weeds

Though aquaculture is practiced mainly based on vertebrate and invertebrate animal species, plants also contribute substantially for aquaculture production. In several countries, aquatic weeds are cultured for human consumption apart from using as fodder and in the manufacture of alginates, mannitol and iodine. Several countries are producing aquatic weeds as raw material for industries. Among the freshwater weeds, *Trapa* sps., *Nasturtium* sps. and *Ipomea* sps. are the important ones and the large scale culture of these weeds is practised in marine waters. The important sea water weeds are *Porphyra* sps. (Red algae), *Undaria* sps. and *Laminaria* sps. (Brown algae), Entermorpha sps. and *Monostroma* sps. (Green algae). The usual culture method is that the monospores are collected on to the suitable substratum and they are transferred to suitable

INTEGRATED FISH FARMING WITH LIVE STOCK

Integrated systems are ideal for location rural areas easier for farmers to understand and operate. Combined culture systems are energy efficient food production systems and by recyling organic wastes, valuable resource in the form of animal protein can be recovered by its production is enhanced from the unit area.

Fish cum Duck culture

It is highly profitable as it greatly enhances the animal protein production in terms of fish and duck per unit area. Ducks are known as living manuring machines. The duck droplings contain 25% organic and 20% inorganic substances with a number of elements such as carbon, phosphorus, potassium, nitrogen, calcium, etc. Hence, it forms a very good source of fertiliser in fish ponds for the production of fish food organisms. Besides manuring, the ducks eradicate the unwanted insects, snails and their larvae which may be vectors of fish pathogenic organisms and water-borne disease causing organisms infecting human beings. For duck-fish culture, the ducks may be periodically allowed to range freely or may be put in screened resting places above the water. About 15-20 day old ducklings are stocked in the sheds. The number of ducks may be between 100 and 3,000 /ha depending on the duration of fish culture and manure requirements. For culturing fish with ducks, it is advisable to release fish fingerlings of more than 10 cm in size. Under stocking density of 20,000/ha and culture period of 90 days, a fish production of 2000 kg/ha has been obtained in duck-fish culture.

Fish cum Poultry culture

Utilisation of poultry manure in fish culture is a recent practice. The droppings of chicks are rich in nitrogen and phosphorous would fertilise the ponds. It has been estimated that 40 laying birds yield about one ton of manure in an year. Manuring the pond with poultry dropplings is found to produce dense population of both phyto- and zooplankton dominated by rotifers. Peak plankton production is observed on 12th day after manure application. Poultry manure is a complete fertiliser which is characteristic of both organic and chemical fertilisers.

Fish cum Pig culture

Pig dung contains more than 70% digestible feed for fish. It is reported that the feed while passing through the alimentary canal of pig, gets mixed with certain enzymes which continue to act even after defecation. The undigested solids present in the pig dung also serve as direct food source of Tilapia and Common carp. A density of 60-100 pigs has been found to be enough to fertilise a fish pond of one hectare area. The optimum dose of pig manure per ha has been estimated may be obtained from 50 pigs. If the manure is to be obtained from a dosage of 400 kg/ha/day for 12 times in a year will be required. Fish like grass carp, silver carp and common carp (1:2:1) are suitable for integration with pigs.

Fish cum Cattle culture

Similar to pig sheds, cow sheds may also be constructed in the vicinity of fish ponds and the cow dung or slurry from the bio-gas plants may be discharged into the fish ponds.

9

NUTRITION AND FEEDS

The knowledge relating to basic nutritional requirements of fish stems from man's endeavours to raise fish for food and stocking in lakes and rivers. In intensive fish culture practices the objective is to maintain optimum density of fish per unit area of water by adopting techniques relating to polyculture, multiple stocking, stock manipulation etc. Under such conditions of fish culture, the natural protein component of food organisms present in the environment will not meet the needs of the growing fish biomass, there by necesitating supplementation with protein rich needs. The development of artificial feeds mainly depend on the studies relating to basic nutrition and physiology. For economic reasons such investigations should aim to find out the minimun level of protein which satisfies the amino acid requirements of the species for optimum inherent capacity for growth, adequate supplementation by carbohydrates to serve as dietary calories and luxurious supply of vitamins and minerals for necessary stimulation of protein digestion.

FLOW OF FOOD ENERGY

The typical flow chart of energy pathways in fish suggests that much of the energy is utilised for maintenance, activity and specific dynamic action which is nothing but the energy expected in digestion, absorption, transportation and excretion. If the feeds lack adequate protein and growth parameters like vitamins and minerals, the food conversion ratio (FCR) would be poor.

$$FCR = \frac{\text{Dry weight of feed fed}}{\text{Wet weight gain}}$$

Hence for better food conversion ratios, a complete and ideal food is required in order to bring down the cost of aquaculture production.

BIOENERGETICS

Bioenergetics is the study of the balance between energy intake in the form of food and energy utilization by animals for life-sustaining processes such as maintenance, activity and tissue synthesis. The original source of food energy is the Sun through photosynthesis, chloroplasts in green plants capture radiant energy from the Sun and convert it into chemical energy through the syntheis of gluscose. This compound serves as the hydrocarbon source from which plants synthesize other organic compounds, primarily carbohydrates, proteins and lipids, which are the primary energy sources for fish and other animals.

The basic unit of heat energy is the calorie (Cal), defined as the amount of heat required to raise the temperature of l g of water 1°C, measured from 14.5 to 15.5°C. This unit is too small for most convenient use in nutrition. Thus the kilocalorie (K Cal) or 1,000 calories is more commonly used. The international unit of work and energy, the joule is also used: 1 Joule = 0.239 calories or 1 calorie = 4.184 joules.

GROSS ENERGY

Energy content of a substance is determined by completely oxidising the compound to carbondioxide, water and other gases and measuring the heat released, which is called the gross energy of the product. This is done with an instrument called an adiabatic bomb calorimeter. Gross energy values for several pure compounds and feed stuffs are presented in Table 9.1.

TABLE. -9.1 : GROSS ENERGY VALUES FOR SOURCES OF CARBOHYDRATES, FATS AND PROTEINS DETERMINED BY BOMB CALORIMETER.

Substrate	*K. Cal/gram*
Glucose	3.77
Corn starch	4.21
Triglyceride beef fat	9.44
Soyabean oil	9.28
Casein	5.84

Note that fats (triglycerids) have approximately twice as much gross energy as carbohydrates. This is related to the relative contents of oxygen, hydrogen and carbon in the compounds. Heat is released only when hydrogen or carbon can react with oxygen from outside of the molecule. Glucose, for example, has enough endogenous oxygen to react with all of the hydrogen in the molecule; therefore, only carbon is oxidized by exogenous oxygen. Because oxidation of a gram of carbon produces only approximately one-fourth as much as heat as oxidation of a gram of hydrogen, fats which have much less endogenous oxygen than carbohydrates, yield more heat upon oxidation than carbohydrates.

AVAILABLE ENERGY

Gross energy content of a food is not a measure of its energy value to the consuming animal. Difference between gross energy and the energy available to the animal for productive purposes varies widely among food materials. Digestibility accounts for most of the differences in available energy among feed stuffs for fish.

Digestible energy is the difference between the gross energy of the food consumed and the energy lost in the feces. Digestible energy in fish can be determined directly or indirectly. In the direct method, total food consumed and total feces excreted are measured; the indirect method involves collecting only a sample of the food and feces and digestion coefficients are calculated on the basis of ratios of energy to indicator in the food and feces. An indicator is an inert, indigestible compound in the food; it may be a natural component such as ash or fibre or it may be an added component such as chronic oxide. Gross energy values for sources of carbohydrates, fats, and proteins are determined by bomb calorimeter.

A major consideration in determining digestion coefficients with fish is collecting feces to avoid losses in the water. This means rapid removal of feces from the water before leaching occurs or removal of feces directly from the digestive tract. More detail on methods used in digestion trails with fish are presented by Choslinger and Bayley (1982) and Lovell and Stickney (1977). Equations below is used to calculate digestible energy (DE) by the total collection method and to calculate DE by the indirect method.

$$\% \text{ DE} = \frac{\text{Food energy - Feces energy}}{\text{Food Energy}} \times 10$$

$$\% \text{ DE} = 100 \ \frac{\text{Energy in food}}{\text{Energy in feces}} \times \frac{\text{Indicator in feces}}{\text{Indicator in food}} \times 100$$

Metabolizable energy, which represents digestible energy less energy lost from the body through gill and urinary wastes, is more difficult to determine. The fish must be confined in a metabolism chamber to collect gill and urine wastes are collected. Metabolizable energy (ME) is calculated according to equation

$$\% \text{ ME} = \frac{\text{Food energy - (Energy lost in feces, urine, gills)}}{\text{Food Energy}}$$

Use of metabolizable energy instead of digestible energy to evaluate fish feeds would allow a more absolute estimate of the dietary energy metabolized by the tissues of the animal; also, the National Research Council Committee on Animal Nutrition has adopted this system. Practically, however, metabolizable energy offers little advantage over digestible energy in evaluating useful energy in feeds for fish because energy loss in digestion accounts for most of the variation in recoverable energy among foods. Energy losses through gill and urinary excretion by fish do not vary among foods nearly as much as fecal energy losses and are smaller than non fecal energy losses by mammals and birds. Further more, confinement of the fish in metabolism chambers to determine metabolizable energy is difficult and stresses the fish. Digestible energy is easier to determine and the fish are not stressed when allowed to feed voluntarily.

ENERGY BALANCE IN FISH

Energy flow in fish is illustrated in Fig. 9.1.

Energy partitioning in fish is similar to that in mammals and birds, but there are several quantitative differences that make fish more energy efficient, especially in the assimilation of protein rich feed stuffs.

Energy losses in urine and gill excretions are lower in fish because approximately 85% of the nitrogenous waste is excreted as ammonia

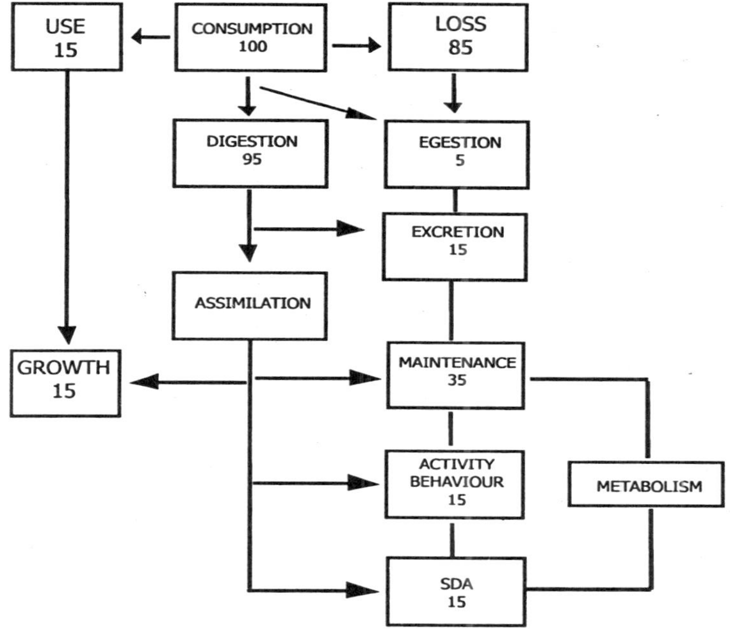

Fig. 9.1: Energy flow in fish

instead of urea (mammals) or uric acid (birds). Heat increment, the rise in energy expenditure associated with the assimilation of ingested food, is lower in fish. Smith, Ramsey and Scott (1978) measured heat increment by direct calorimetry and found it to be 3% to 5% of ME for rainbow trout; in mammals it may account for as much as 30% of ME. Most of this difference in heat increment is caused by the relatively high energy cost of synthesizing and excreting urea and uric acid as compared with ammonia.

NUTRITIONAL REQUIREMENTS

Fast growing varieties of fish are mostly chosen for culture practices. The general practice is to provide some starchy foods to several fast growing fish to serve as dietary calories. Due to much research in the field of fish nutrition lead to the formulation of several balanced artificial feeds. To meet the dietary demand of fish one should know the nutritional requirements of fish such as proteins, carbohydrates, fats, vitamins, micronutrients, etc., besides the knowledge relating to digestibility and utilisation of the compounded feeds by the fish for yielding protein as the final metabolised product in the intensive fish culture practices.

CARBOHYDRATES

After the proteins and lipids, the carbohydrates represent the third most abundant group of organic compounds in the animal body. By contrast, carbohydrates constitute the major class of organic nutrients within plant tissues. The carbohydrate group includes such important compounds as glucose, fructose, sucrose, lactose, starch, glycogen, chitin and cellulose.

Carbohydrates are the most abundant and relatively least expensive source of energy in animal aquaculture. These may consist of easily digested sugars to complex cellulose which is difficult to digest. Most of the cultivable fishes like carps and mullets are omnivorous taking in considerable amount of vegetable mayyet and are therefore, well adapted physiologically to digest starch. Digestibility of starch is reported to be 30-90%. Rice bran and wheat bran which are the main starchy diets used for cultivable fishes are found to be highly digestible.

Carbohydrates are usually defined as substances containing carbon, hydrogen and oxygen, with the last two elements being present in the same ratio as in water [i.e.,$C_x(H_2O)_y$]. Although this definition is satisfactory for the majority of the compounds present within this group, a few carbohydrates contain a lower proportion of oxygen than in water or exist as carbohydrate derivatives which may contain nitrogen and sulphur.

Classification

The carbohydrates can be divided into two major groups according to their chemical structure; the sugars and non-sugars. The simplest sugars are called monosaccharides, and these in turn can be divided into five sub-groups depending on the number of carbon atoms present in the molecule: Trioses ($C_3H_6O_3$), Tetroses ($C_4H_8O_4$), Pentoses ($C_5H_{10}O_5$), and Hexoses ($C_6H_{12}O_6$). These monosaccharides may also in turn be linked together (with elimination of water) to form di, tri or polysaccharides containing two, three or more monosaccharides units or residues respectively. Here the term 'sugar' is restricted to those carbohydrates containing less than 10 monosaccharide units. Non-sugars are therefore carbohydrates which contain more than 10 monosaccharide units and which do not possess a sweet taste. The non-sugars can be divided into two sub-groups, homopolysaccharides and heteropolysacharides; the former consisting of identical monosaccharide units and the latter of mixtures of different monosaccharide units.

Monosaccharides

In general all monosaccharides are soluble in water, sparingly soluble in ethanol and insoluble in ether, are optically active, possess reducing properties (i.e., reduce Fehling's solution) can be represented by the general formula $C_xH_{2x}O_x$, and generally a sweet taste.

The chain formula of some of the more common monosaccharides can be represented thus:

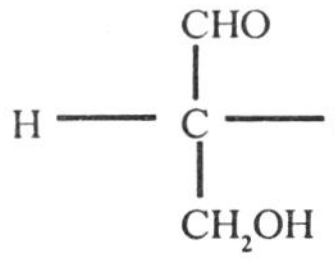

CHO
|
H — C — OH
|
CH_2OH

Triose (C_3 H_6O_3)

D (+) – Glyceraldehydc

CHO
|
H — C — OH
H — C — OH
H — C — OH
|
CH_2OH

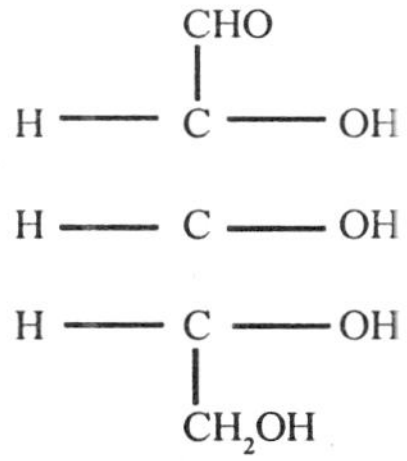

Pentose (C_5 $H_{10}O_5$)

D (+) Ribose

Hexose ($C_6H_{12}O_6$)

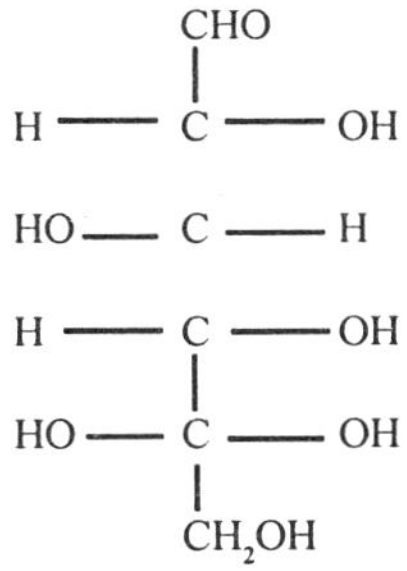

CHO
|
H — C — OH
HO — C — H
H — C — OH
|
HO — C — OH
|
CH_2OH

D (+) Glucose

Hexose ($C_6H_{12}O_6$)

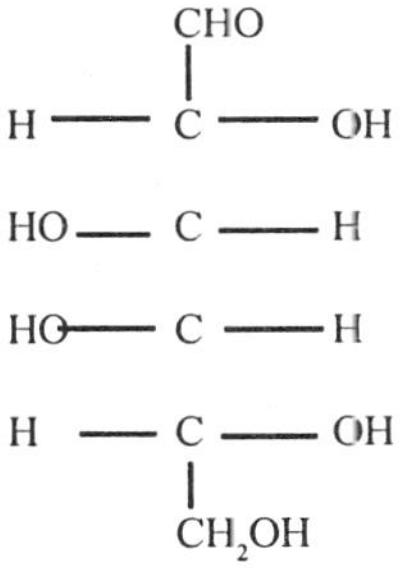

CHO
|
H — C — OH
HO — C — H
HO — C — H
H — C — OH
|
CH_2OH

D (+) Galactose

Within the structural formulae presented 'D' represents the configuration or direction of the hydroxyl group (OH) on the carbon atom next to the last carbon from the functional or aldehyde group. For example, in the case of D(+)-Glyceraldehyde and D(+)-Glucose the hydroxyl group on the penultimate carbon atom (i.e., C_2 and C_5 respectively) is on the right when the aldehyde group (CHO) is at the top of the formula. Similarly, the symbol (+) or (-) denotes the direction of the optical rotation observed

Hexose ($C_6H_{12}O_6$)

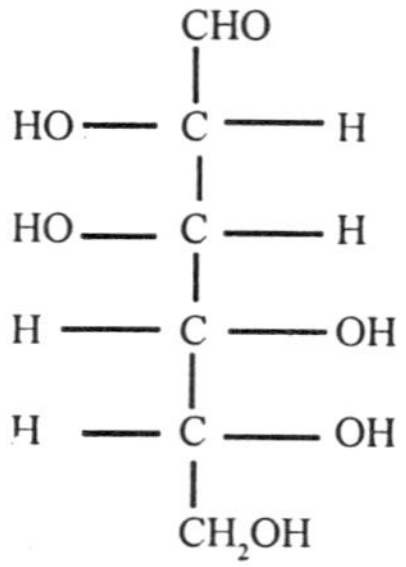

D (+) Mannose

Hexose ($C_6H_{12}O_6$)

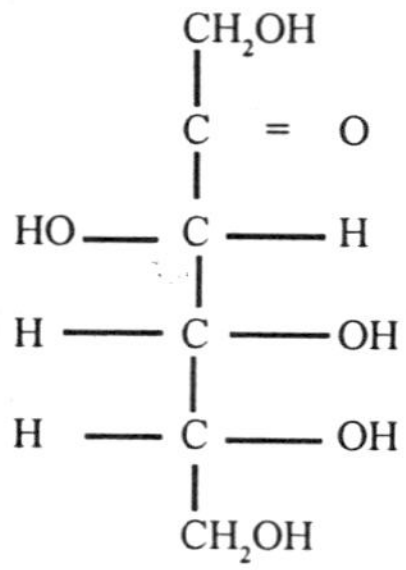

D (-) – Fructose

when a solution of that sugar is placed in a polarimeter; dextro-rotatory (clockwise direction +) or laevorotatory (anticlockwise direction, (-). Virtually all naturally occurring monosaccharides are members of the D series of sugars, the configuration around the penultimate carbon atom being the same as that in D-glyceraldehyde. In addition, all naturally occurring monosaccharides are dextro-rotatory, with the exception of fructose and erythrose.

There is considerable evidence to suggest that monosaccharides may also exist in a ring or cyclic molecular form. For example, two cyclic forms of D-glucose are known to occur in nature α-D-glucose and β-D-glucose. As with the main chain formulae, the difference between these two cyclic forms depends upon the configuration or direction of the hydroxyl group on carbon atom 1.

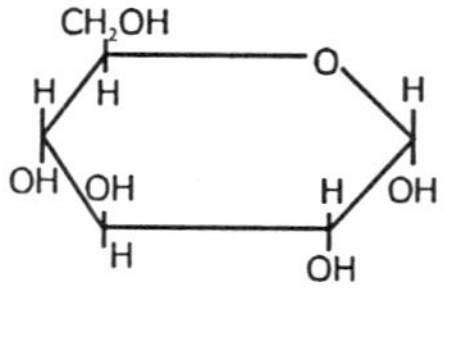

α - D- Glucose

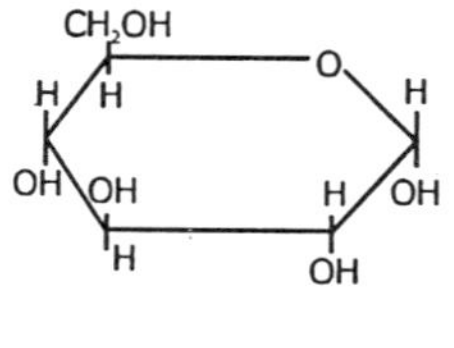

β- D- Glucose

The biological importance of the structural difference between αand -β-glucose must be stressed here; the structural configuration determining the physical and subsequent biological properties of polysaccharide composed of individual monosaccharide units. For example, the polysaccharides cellulose is composed of insoluble zig-zag chains of β-glucose units, whereas the polysaccharides starch and glycogen are composed of more biological reactive helical or branched chains of α-glucose units.

It should be mentoned here at this stage that monosaccharides are seldom directly involved in biochemical reactions within the cell, but are first transformed into appropriate monosaccharide derivatives. Important monosaccharide derivatives include the sugar phosphate esters (D-glucose-6-phosphate, D-glucose-1-phosphate, D-fructose-6-phosphate and phosphate diesters), amino sugars (D-glucosamine), sugar acids (gluconic acids, glucoronic acid) and sugar alcohols (sorbitol).

D- Glucose-6-Phoshate β- Glucoronic acid α- D-Glucosamine

Pentoses

Important pentose monophosphates include L-arabinose, D-xylose and D-ribose. From a nutritional viewpoint the most important pentose is D-ribose and its derivatives D-deoxyribose and ribitol. For example, D-ribose and D-deoxyribose are essential of ribonucleic acid (RNA) and deoxyribonucleic acid (DNA) respectively, and ribitol an essential component of the vitamin riboflavin.

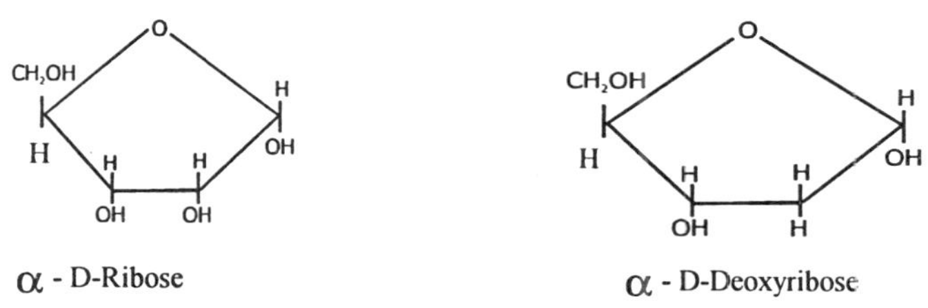

α - D-Ribose α - D-Deoxyribose

HEXOSES

Glucose

Occurs in its free form in plant tissues, fruits, honey and blood. In most natural food stuffs glucose is present in a combined form, either as the sole monosaccharide component of disaccharides (i.e. maltose) and polysaccharides (i.e. starch, glycogen, cellulose) or combined with other monosaccharides in the form of lactose (milk sugar), sucrose and heteropolysaccharides. Glucose is fermented by yeast during the manufacture of beers and wine to yield alcohol and carbon dioxide. Similarly as fruit ripen their free sugar or glucose content normally increases and starch content decreases accordingly.

Fructose

Like glucose, fructose is found in its free state in plant juices, fruit and honey. It is a component of the disaccharide source, and is the sweetest sugar known in the nature (i.e., it is responsible for the exceptional sweet taste of honey).

Galactose

Although galactose does not occur in its free state in nature, it exists as a component of the disaccharide lactose and many polysaccharides, including galatolipids, gums and mucilages.

Disaccharides

Disaccharides are composed of two hexose sugars joined together with the elimination of water, thus:

$$C_6H_{12}O_6 + C_6H_{12}O_6 = C_{12}H_{22}O_{12} + H_2O$$

The most important naturally occurring disaccharides are maltose, sucrose and lactose.

Maltose is composed of two molecules of glucose joined together through a α -1,4-glycoside linkage. Maltose is a reducing sigar and is soluble in water. Maltose is not found in nature to any extent but is a product obtained during the enzymatic degradation of starch. For example, maltose is produced from starch during the germination of barley by the action of the enzyme amylase; after germination and drying the barley or malt is used in the manufacture of beer andd malt whisky.

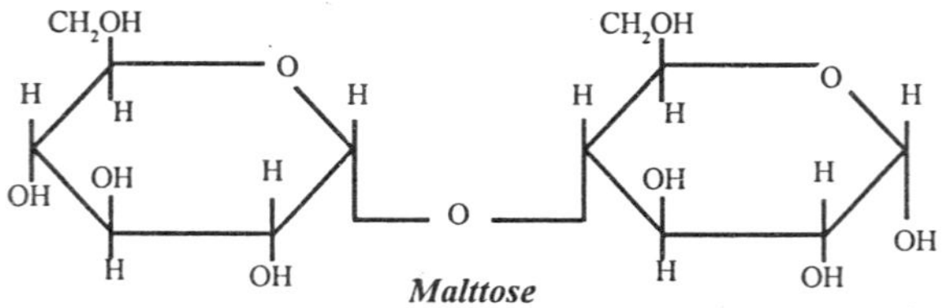

Malttose

Sucrose is composed of one molecule of glucose and fructose joined together through a α-1, β -2-glycoside linkage. Since both functional reducing groups are involved in the glycoside linkage, sucrose does not possess reducing properties. Sucrose is widely distributed in nature and occurs in most plants; rich sources of sucrose include sugar cane (20% sucrose), sugar beet (15-20%), mangolds and carrots. Sucrose is the sugar of familiar use in the domestic household. When sucrose is heated to a temperature of 160° C it forms barley sugar and at 200° C forms caramel. Beet or cane molasses are agricultural by-products obtained from the manufacture of sucrose from sugar beets or sugar cane respectively. Molasses are dark coloured, viscous liquids (20-30% moisture), from which no further sucrose can be separated by usual crystallization procedures because of the presence of appreciable quantites of reducing sugars (i.e., glucose) and non-sugar impurities.

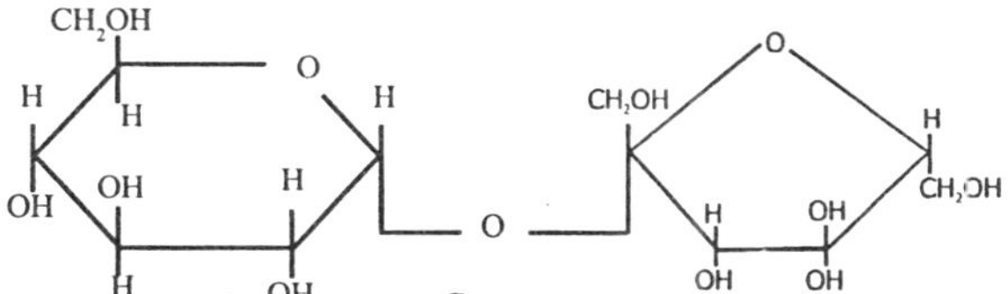

Sucrose

Lactose is composed of one molecule of glucose and galactose joined together by an -1, 4-glycoside linkage. Like maltose it has reducing properties. Lactose or milk sugar, is the principal sugar found in the milk and is unique to mammals. It forms about 40% of the total milk solids; the total lactose content of cows milk and human milk being 4.6-4.8% and 7% respectively. Lactose readily undergoes bacterial fermentation; for example, the souring of milk by *Streptococcus lactus* by the fermentation of lactose to lactic acid. Like sucrose, if lactose is heated to temperatue of 175° C it forms lactocaramel.

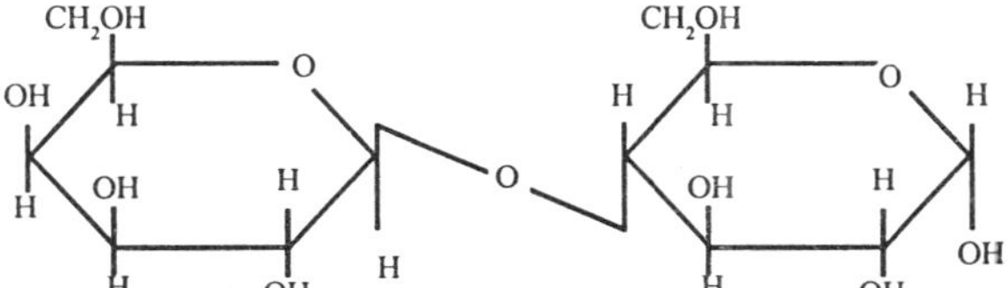

Lactose

Homopolysaccharides

These carbohydrates are very different from the sugars. They have a high molecular weight and are composed of large numbers of hexose or to a lesser extent pentose residues. Many of them occur in plants or animals as a reserve food material (i.e., starch or glycogen) or as structural materials (i.e., cellulose or chitin).

Starch is composed of two structural components, amylose and amylopectin. Although the relative proportions of amylose and amylopectin within plant starches varies depending on the species (20-30% amylose and 70-80% amylopectin), the fundamental unit of these two structural components is α-D-glucose. For example, amylose consists of long unbranched chains of 100 or more D-glucose units joined together by α-1, 4 linkages. On the otherhand, amylopectin is composed of highly branched chains of D-glucose units (20-30 units per branch); the units being joined together by α-1, 4 linkages and also α-1, 6 linkages (the α-1, 6 glycoside linkages used at the beginning of a side chain).

Starch is the depot form of sugar or glucose in plants; it occurs in stems, fruits, seeds, roots or leaves, forming the greatest carbohydrate food reserve of plants and consequently constitutes the largest carbohydrate component of animal feeds. For example, starch may account for up to 70% by weight of seeds and up to 30% of fruit, tubers or roots. Starch is stored within plants in the form of granules, the size and shape of which varies from species to species. Each granule is surrounded by a thin layer of cellulose which render them insoluble in water and undigestible in their raw or cooked form by non-ruminant animals including fish and shrimp. Cooking, by heating in the presence of moisture will however facilitate the rupture of the cellulose membrane, resulting in the absorption of water by the starch within, which in the presence of heat will result in the gelatinization of the starch forming a gelatinous solution or paste. When starch is subjected to dry heat, dextrin is formed; dextrin being an intermediate degradation product of starch in the sequence Starch → dextrin → maltose → glucose. For example, starch in bread is converted to dextrin when toasted, the dextrin giving toast its characteristic taste.

Glycogen is composed of branched chains of -D-glucose units together by β-1,4 linkages and α-1, 6 linkages; the latter being more numerous in glycogen (as compared with amylopectin) due to the presence of more and shorter branches of 10-20 glucose units. Glycogen is the form in which carbohydrate is stored within the animal body; being particularly concentrated in the liver and muscle.

Cellulose is composed of very long chains of D-glucose units joined together by α-1, 4 linkages. It is a very stable polysaccharide and is most abundant carbohydrate in nature; forming the fundamental structure of the plant cell wall. Cellulose has great tensile strength and is resistant to chemical attack. Although cellulose can be hydrolysed by a strong acid treatment, with the exception of micro-organisms, few non-ruminant animals have the necessary endogenous enzymes (i.e., cellulses) capable of hydrolysing and digesting cellulose. For example, the cellulase enzymes which are capable of attacking cellulose are only found in germinating seeds, fungi, and bacteria (i.e., such as those present in the digestive tract of ruminants). An example of nearly pure form of cellulose is cotton.

Chitin is composed of repeating units of N-acety l-D-glucosamine joined together by β -1, 4 linkages and is therefore similar in structure to cellulose. Chitin is the major structural component of the cuticle of insects and the exoskeleton of crustaceans.

Heteroploysaccharides

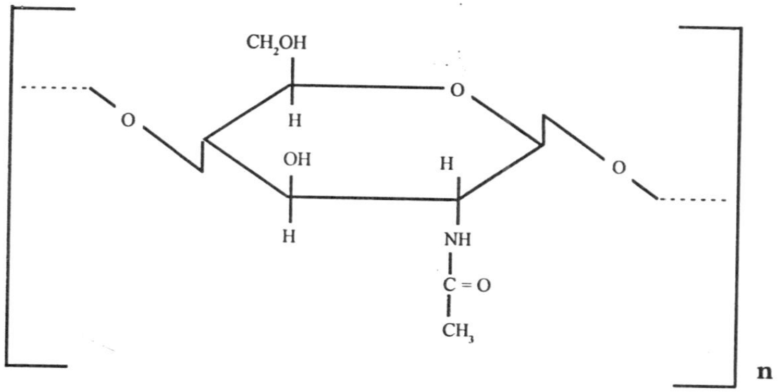

Chitin

In contrast to the homopolysaccharides, the heteropolysaccharides consist of high molecular weight mixtures of different monosaccharide units.

Hemicellulose is composed principally of xylose units linked together by β-1, 4 linkages, but may also contain hexoses and sugar acids (i.e., uronic acid). These polysaccharides normally accompany cellulose in the leafy and woody structure of higher plants and seeds. They are insoluble in water and like cellulose are not digested by non-ruminant animals.

Gums are constituents of plant wounds and are very complex compounds; on hydrolysis yielding a variety of monosaccharides and sugar acids. An example of gum is gum arabic (Acacia gum).

Mucillages complex carbohydrates present in certain plants and seeds. Many algae and especially seaweeds, yield mucillage which are soluble in hot water and which form a gel on cooling. Agar, as ulphuric acid polymer of galactose, is a widely used mucillage or gel obtained from seaweeds (Gelidium family). Other examples include alginic acid derived from brown seaweeds (Laminaria family).

Pectic substances are complex carbohydrates which contain D-galacto-uronic acid as the main constituent.

Mucopolysaccharides are complex carbohydrates which contain sugars and uronic acids, and constitute the mucous secretions of animals. They are acidic nature and may be rich in sulphate ester groups.

Carbohydrate Function

Carbohydrates are synthesized in all green plants by a process called photosynthesis, which can be represented thus;

$$6CO_2 + 6H_2O + \underset{(673\ \text{kcal})}{\text{Light}} \longrightarrow C_6H_{12}O_6 + {}^{6}O_2$$

Within man and terrestrial farm animals dietary carbohydrates serve as the principal source of metabolic energy (ATP). This reaction can be represented as:

$$C_6H_{12}O_6 + 6O_2 \longrightarrow 6CO_2 + 6H_2O + 38\ \text{ATP}$$

Although carbohydrates may be regarded as non-essential dietary nutrients for fish and shrimp, their inclusion in practical diets is warranted because:

— They represent an inexpensive source of valuable dietary energy for non-carnivorous fish and shrimp species.

— Their careful use in practical diets can spare the more valuable protein for growth instead of energy provision (a procedure called "protein sparing").

They serve as essential dietary constituents for the manufacture of water stable diets when used as binders.

Certain carbohydrate sources serve as dietary components which can increase feed palatibility and reduce the dust content of finished feeds.

The endosperm of grains is composed mostly of starch and this is the major source in animal feeding. Two forms of starch are found in the starch granules in grains; amylose and amylopectin. Amylose is a straight chain α-1/4 - glucose polymer. It comprises 20% to 30% of the starch granule and is soluble in warm water. Amylopectin is a branched chain glucose polymer; the 1-4 straight chain is branched by 1-6 linkage to a side chain. Starch granules are different in morphology and solubility among plant species. Some are quite resistant to rupture, which is necesary for digestion. Most heating ruptures or gelatinizes, the granule and increases solubility and digestibility of the starch. Glycogen is the carbohydrate energy reserve in animal tissue, mainly the liver. It is similar to amylopectin in molecular weight, but its 1-6 linked side chains are shorter and it is more soluble in water.

Carbohydrates may be divided into two major groups according to their chemical structure: the sugars and non-sugars.

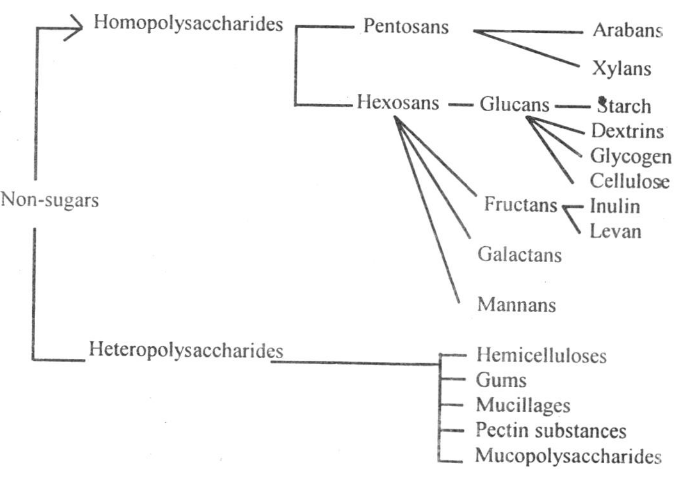

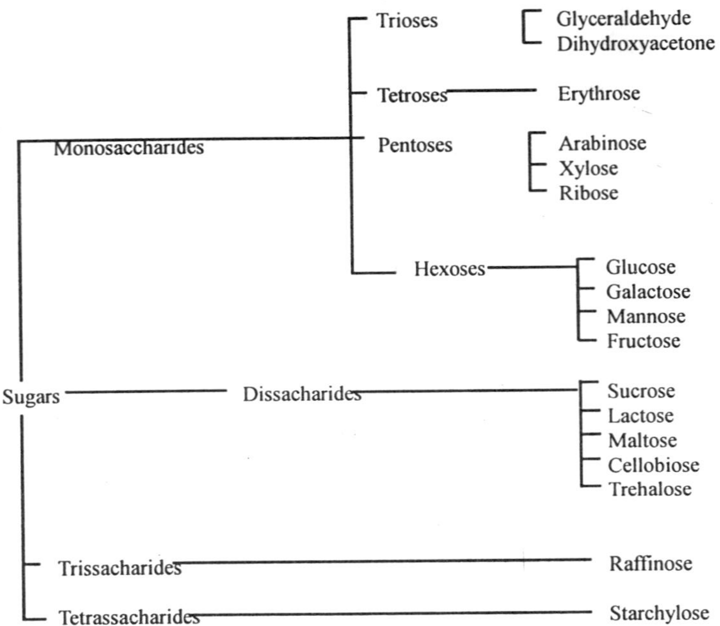

Although carbohydrates are significant source of energy and are components of a number of body metabolites, such as blood glucose, nucleotides and glycoproteins, they are not essential nutrients. The ratio of protein to carbohydrate in the feeding of 1:7 or 1:8 which gives a wide scope to utilize feeding of cheap carbohydrate diets as long as for protein in the natural food is sufficient for growth. While formulating the balance diets, carbohydrate and protein ratio needs a careful manipulation so as to spare the proteins for growth and carbohydrate to serve supplying the dietary calories.

PROTEINS AND AMINO ACIDS

The proteins are among the most important constituents of all living cells and represent the largest chemical group in the animal body, with the exception of water; the whole fish carcass contains on average 75% water, 16% protein, 6% lipid and 3% ash. Proteins are essential components of both the cell nucleus and cell protoplasm and acordingly account for the bulk of the muscle tissues, internal organs, brain, nerves and skin.

Composition

Proteins are very complex organic compounds of high molecular weight. In common with carbohydrates and lipids, they contain carbon (C), hydrogen (H), and oxygen (O), but in addition also contain about 16% nitrogen (N:range 12-19%) and sometimes phosphorus (P) and sulphur (S).

Structure

Proteins differ from other biologically important macromolecules such as carbohydrates and lipids in their basic structure. For example, in contrast to the basic structure of carbohydrates and lipids, which is often composed of identical or very similar repeating units (i.e., the glucose repeating unit within starch, glycogen and cellulose), proteins may have up to 100 different basic units (amino acids). It follows therefore that greater compound variabilities and ranges are possible, not only to composition, but also to protein shape.

Chemical structure

Colliodal in nature, proteins vary in their solubility in water, from insoluble keratin to the highly soluble albumins. All proteins can be denatured by heat, strong acid, alkali, alcohol, acetone, urea and by heavy metals, salts. When proteins are denatured they loose their unique structure and therefore possess different chemical, physical and biological properties (i.e., inactivation of enzymes by heat).

Classification

Proteins may be classified into three main groups according to their shape, solubility and chemical composition.

(a) Fibrous proteins

Insoluble animal proteins which are generally very resistant to digestive enzyme breakdown. Fibrous proteins exist as elongated filamentous chains. Examples of fibrous proteins include the collagen (main protein of connective tissue), elastin (present in elastic tissues such as arteries and tendons), and keratin (present in hair, wool and hooves of mammals).

(b) Globular proteins

Include all enzymes, antigens and hormone proteins. Globular proteins can be further subdivided into albumins (water soluble, heat-coagulable proteins which occur in eggs, milk, blood and many plants); globulins (insoluble or sparingly soluble in water, and present in eggs, milk, and blood and serve as the main protein reserve in plant seeds); and histones (basic proteins of low molecular weight, water soluble, occur in the cell nucleus associated with deoxyribonucleic acid (DNA).

(c) Conjugated proteins

These are proteins which yield non-protein groups as well as amino acids on hydrolysis. Examples include the phosphoproteins (casein of milk, phosviton of egg yolk), glycoproteins (mucous secretions), lipoproteins (cell membranes), chromoproteins (haemoglobin, haemocyanin, cytochrome, flavoproteins) and nucleoproteins (combination of proteins with nucleic acids present in the cell nucleus).

Protein function

The function of proteins may be summarised as follows:

— To repair worn or wasted tissue (tissue repair and maintenance) and to rebuild new tissue (as new protein and growth).

— Dietary protein may be catabolised as a source of energy or may serve as a substrate for the formation of tissue carbohydrates or lipids.

— Dietary protein is required within the animal body for the formation of hormones, enzymes and a wide variety of other bilogically important substances such as antibiotics and haemoglobin.

Protein requirements

The study of dietary nutrient requirements in fish and prawn has been almost entirely based on studies comparable to those conducted with terrestrial farm animals. It follows therefore that almost all the available information on the dietary nutrient requirements of aquaculture species is derived from laboratory based feeding trails; the animals being kept in a controlled environment at high density and having no access to natural food organisms.

Optimum dietary protein level

Based on feeding techniques pioneered and developed for terrestrial animals the dietary protein requirements of fish were first investigated in the Chinnok salmon, *Oncorhynchus tshawytscha* by Delong *et al.* (1958). Fish were fed a balanced diet containing graded levels of high quality protein (casein:gelatin mixture supplemented with crystalline amino acids to simulate the aminoacid profile of whole hen's egg protein) over a 10 week period and the observed protein level giving optimum growth was taken as the requirment (Fig. 9.2).

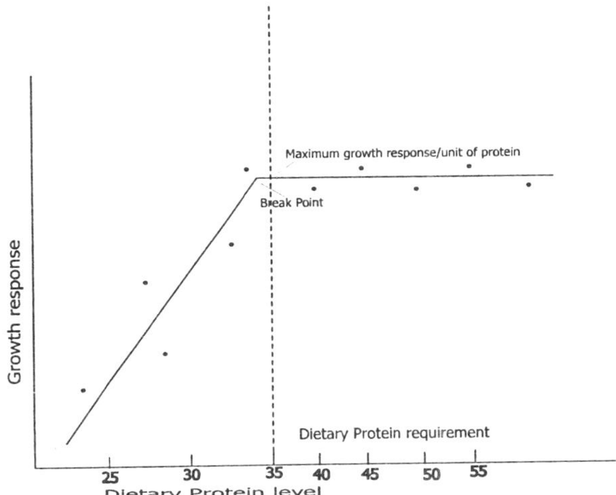

Fig. 9.2 Typical dose response curve

The dietary protein requirements are normally expressed in terms of fixed dietary percentages or as a ratio of protein to dietary energy. The high dietary protein requirement of fish and shrimp is generally attributed to their carnivorous/omnivorous feeding habit, and their preferential use of protein over carbohydrates as a dietary energy source. In contrast to terrestrial farm animals fish and shrimp are able to derive more metabolisable energy from the catabolism of proteins than from carbohydrates.

Dietary protein is the main source of nitrogen and essential amino acids in animals. It is also the most expensive source of energy in artificial diets. In nature, carnivorous fish consume food which are about 50% protein. They have a very efficient system for excretion of waste nitrogen from protein, which is catabolised for energy. Therefore high-protein diets are not harmful but, being expensive, it is necessary to keep the proportion of protein to optimum levels necessary for good growth and feed conversion.

Types of protein found in the fish body are generally based upon function or solubility. Fibrous proteins are highly insoluble (indigestible) proteins and include collagen, elastin and keratin. Collagen is the component of connective tissues, bone matrix, skin, scar tissue, fins, gill operculum, and blood vessels. Elastins are found in arteries, tendons and other stretch tissues. Keratins are found in hair and hooves of land animals, but in only small amounts in fish contractile protein is the muscle protein complex. Three proteins actin, tropomyosin B, and myosin take part in muscle contraction. Muscle protein is highly digestible and high nutritional value. Globular proteins are proteins extractable from tissue with water or dilute salt solutions. They represent enzymes, protein hormones, and proteins of the serum (soluble) fractions of blood.

ESSENTIAL AMINO ACIDS

Although over 100 different amino acids have been isolated from biological materials, only 25 of these are commonly found in proteins. Individual amino acids are characterised by having an acidic carboxy group (-COOH) and a basic nitrogenous group (generally an amino group: -NH_2). In view of the presence of both acidic and basic groups, amino acids are amphoteric (i.e., have both acidic and basic properties) and consequently act as buffers by resisting changes in pH. The chemical and nature of the more commonly occurring amino acids is shown below:

CLASSIFICATION OF AMINO ACIDS

1. Mono amino mono carboxylic acids	a. Glycine b. Alanine c. Valine d. Leucine e. Isoleucine f. Serine g. Threonine
2. Mono amino di-carboxylic acids	a. Aspartic acid b. Glutamic acid
3. Diamino mono carboxylic acid	a. Arginine b. Lysine
4. Sulphur containing amino acids	a. Cysteine b. Methionine c. Cystine
5. Aromatic and heterocyclic amino acids	a. Phenylalanine b. Tyrosine c. Tryptophan d. Histidine e. Proline

Amino acid function

Amino acids occupy a central position in cellular metabolism since almost all biochemical reactions are catabolised by enzymes composed of amino acid residues. Amino acids are essential for carbohydrate and lipid metabolism, for the synthesis of tissue proteins and many important compounds (i.e, adrenalin, thyroxine, melanin, histamine, porphyrins, haemoglobin, pyrimidines, and purines, nucleic acids, choline, folic acid and nicotinic acid, vitamins, taurine, bile salts, etc., and as metabolic source of energy of fuel.

Amino acids can be divided into two nutritional groups, essential and non–essential. The essential amino acids are those that the animal can not synthesize or can not synthesize in sufficient quantity to support maximum growth. The non-essential amino acids are those that can be synthesized by the animal in quantity to support maximum growth. Most monogastric animals, including fish, require the same 10 essential amino acids arginine, histidine, isoleucine, leucine, lysine, methionine, phenylalanine, threonine, tryptophan and valine.

Qualitative requirements for amino acids are determined in fish by feeding a purified diet composed of crystalline amino acids as a control diet and feeding test diets. Similar to the control except that one amino acid at a time has been removed. Test diets that produce no growth or markedly less than the control represent amino acids that are essential to the fish. Quantitative requirements for essential amino acids are determined by feeding graded levels of one amino-acid at a time in a test diet is usually similar to that in chicken or fish eggs, or of the fish muscle. Growth data from the amino acid feeding trails are equated to amount of amino acid in the diet and the requirement is determined by estimating or calculating the break point in the response curve. In early studies involving salmon, the requirement was estimated by visual inspection. Later, studies with channel cat fish used the statistical method of Robbins, Norton and Baker (1979) to determine the break point in the growth response curve. Santiago (1985) determined two requirements for essential amino acids with Nile Tilapia, one for maximum growth (v max) and one for a level of growth less than maximum (Y^1) but within the 95% confidence limit of v max. This is possible by fitting the growth response data with non linear regressions, as described by Zeitoun, Ulbrey and Magee (1976).

Dietary imbalances of amino acids can cause reduced performance by animals through amino antagonism or toxicity. When some amino acids are fed in excess of their required levels, they cause an increase in the requirement for other structurally similar amino acids, or amino acid antagonism. In some instances, however, dietary excess of certain amino acids are directly toxic and their negative effects cannot be ameliorated by additions of other amino acids; this in amino acid toxicity. Fish diets containing practical feed stuffs, such as grain by products, oil meals and animal by products are not likely to be so seriously imbalanced, but special diets could be.

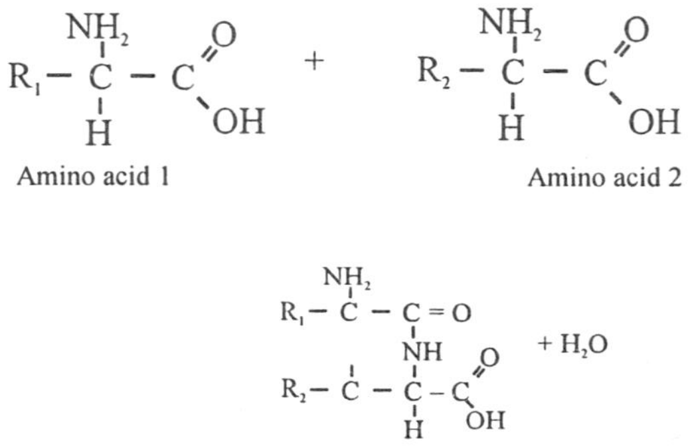

$$R_1-\overset{NH_2}{\underset{H}{C}}-C\overset{\nearrow O}{\searrow OH} \quad + \quad R_2-\overset{NH_2}{\underset{H}{C}}-C\overset{\nearrow O}{\searrow OH}$$

Amino acid 1 Amino acid 2

$$\begin{array}{l} R_1-\overset{NH_2}{C}-C=O \\ \qquad\qquad\; | \\ \qquad\qquad NH \\ \qquad\qquad\; | \\ R_2-C-\underset{H}{C}-C\overset{\nearrow O}{\searrow OH} \end{array} \quad + H_2O$$

Peptide linkage

Fig.: 9.3 Basic Structure of amino acid.

PROTEIN SYNTHESIS IN FISH

Protein synthesis in animal tissue is a complex process which involves deoxyribonucleic acid (DNA), ribonucleic acid (RNA) and ribosomes. DNA, a chromosomal component of cells, carries the genetic information in the cell and transmits inherited characteristics from one generation to the next. The structure of DNA consists of four types of nitrogenous bases (adenine, guanine, cytosine and thymine) linked to a backbone of alternating phosphate and deoxyribose groups. The DNA molecule is in the form of a long double helix, which consists of two chains running in opposite directions and linked together through hydrogen bonds where adenine always pairs with thymine and guanine pairs with cytosine. The sequence of the nitrogenous bases can vary infinitely and this sequence determines the exact protein to be synthesized. DNA controls the development of the organism by controlling the formation of RNA. The composition of RNA is similar to that of DNA, except that ribose is the sugar instead of deoxyribose and uracil replaces thymine. The nucleotides of RNA are linked through their phosphate groups to form long single chains which might fold and form areas of double helical structure.

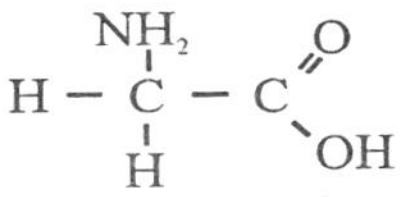

Glycine

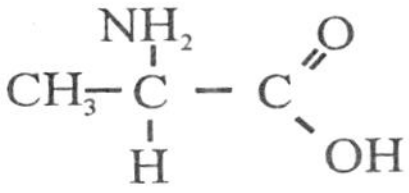

Alanine

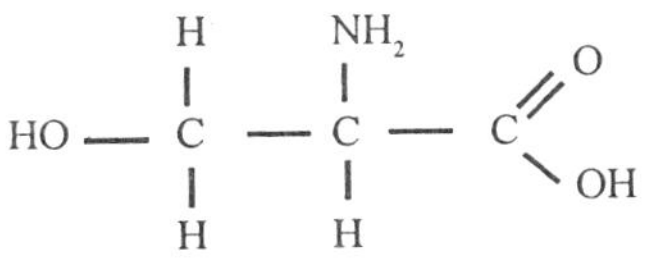

Serine

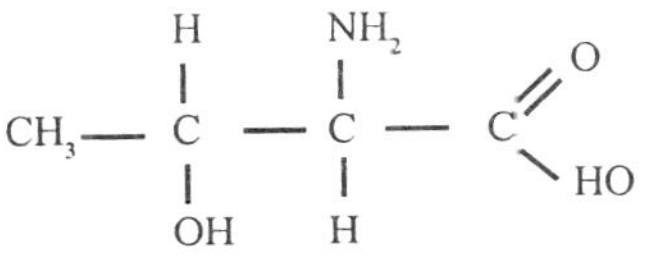

Threonine

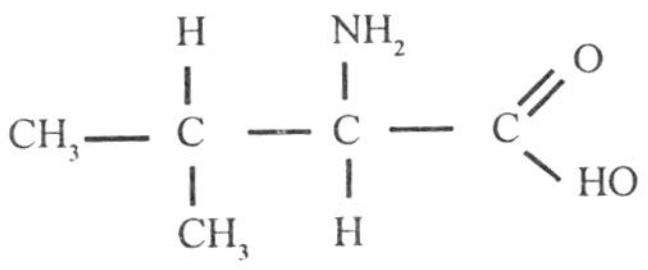

Valine

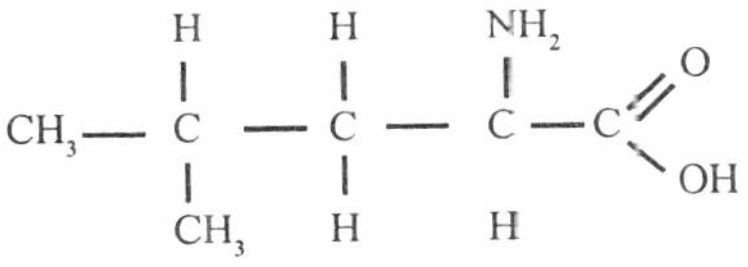

Leucine

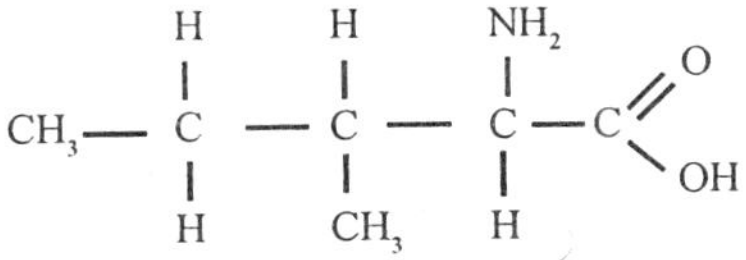

Isoleucine

Fig. 9.4: Aliphatic amino acids
(Mono amino mono carboxylic acids)

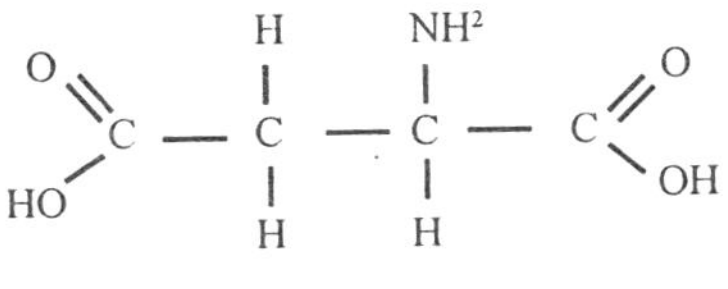

Aspartic acid

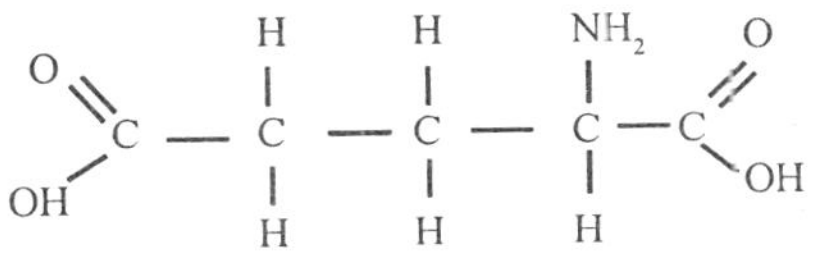

Glutamic acid

Fig. 9.5: Acid amino acids

(Mono amino mono dicarboxylic acids)

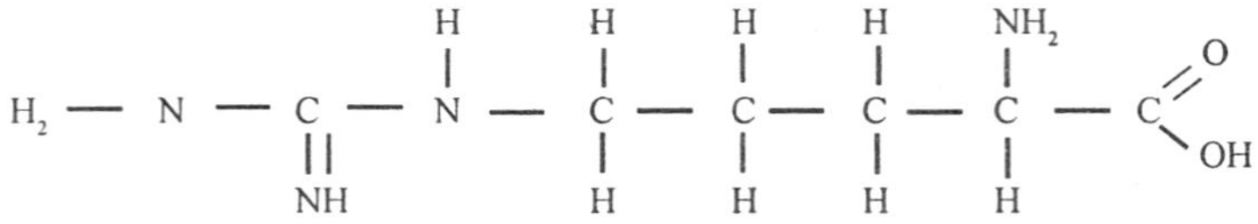

Alginine

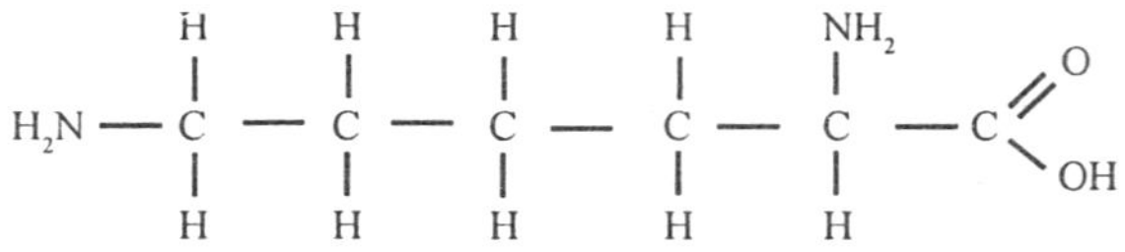

Lysine

Fig. 9.5: Diamino monocarboxgylic acids

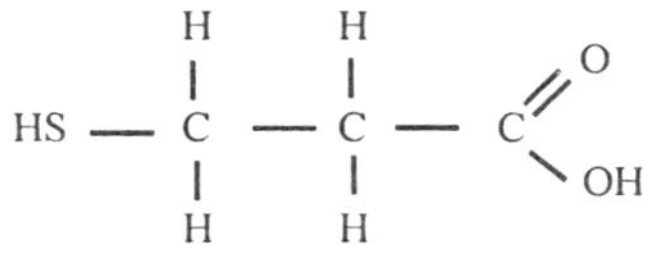

Cysteine

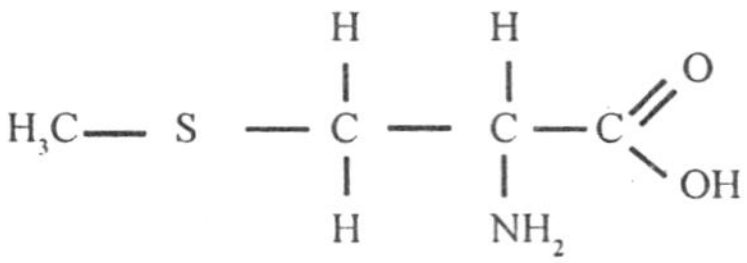

Methionine

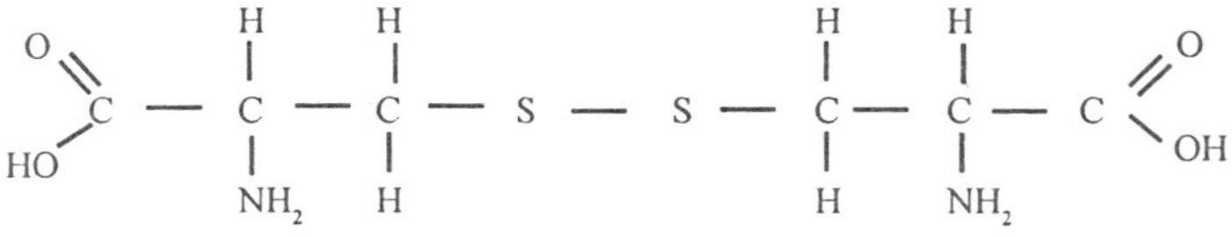

Cystine

Fig. 9.6: Sulphur containing amino acids

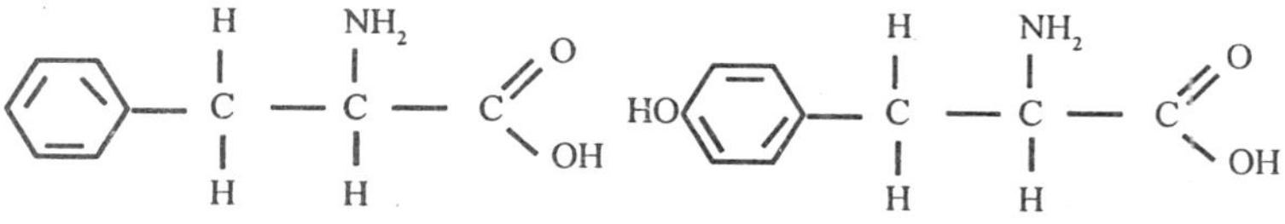

Phenyl alanine

Tyrosine

Fig. Aromatic amino acids

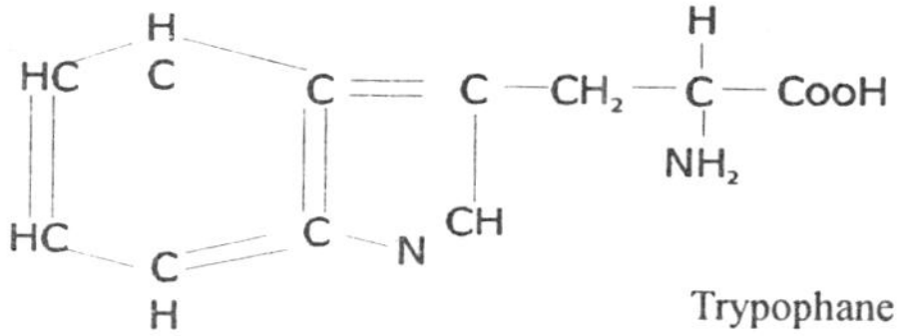

Trypophane

Fig.9.7: Heterocy clic amino acids

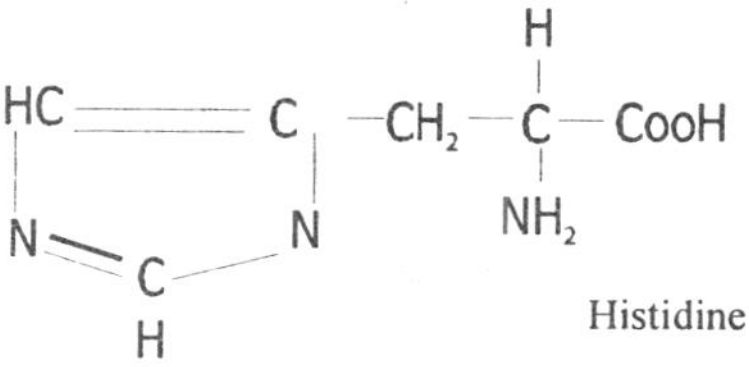

Histidine

Fig.9.7: Heterocyclic amino acids

Proline

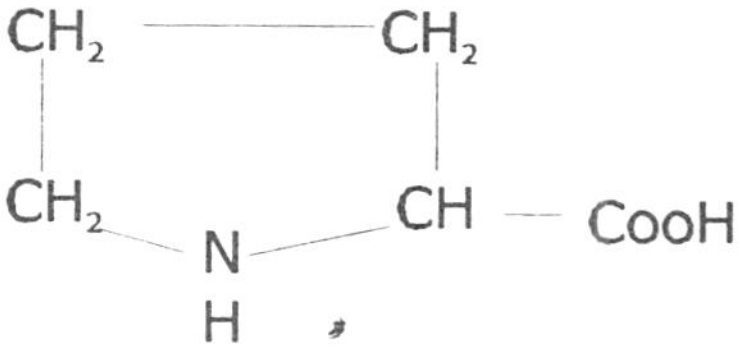

Fig. 9.8: Heterocyclic amino acids

There are three kinds of RNA in cells that participate in protein synthesis; messenger RNA, transfer RNA and ribosomal RNA. Messenger RNA carries the code transcribed from DNA and determines the sequence of amino acids in the protein being formed. Transfer RNA's carry specific amino acids to the ribosomes where they interact with messenger RNA. Ribosomal RNA is part of the structure of the ribosome, which is the site of protein formation in the cell.

Thus, amino acids are linked in sequence predetermined by the sequence of nitrogenous bases in messenger RNA and in turn, in DNA. The addition of amino acids to a growing polypeptide chain occurs very rapidly. For example, synthesis of the protein chains in haemoglobin occurs at the rate of two amino acid additions to the chain per second so that the proteins, which contain 141 to 146 amino acid residues are synthesized in about 1.5 minutes.

The requirements are the highest in the initial feeding of the fry and decrease as fish size increases. For maximum growth, young fish require between 40 to 60% of their diet as proteins. Most of the wet weight gain in the lean fish is in the form of muscle tissue.

The gross protein requirements of a number of fin fish are as follows:

Species	Crude protein level in diet for optimum growth (g/kg)
Rainbow trout	400 - 600
Common carp	380
Salmon	400
Eel	445
Grass carp	410 - 430

Protein requirements are influenced by water, temperature, body size, stocking density, oxygen levels and the presence of toxins. Readily digested high protein materials have higher metabolizable energy (ME) values for fish than other mono-gastric animals. Similarly, proteins has more net energy for fish than it has for mammals or birds. Fish are among the most efficient of all animals in converting feed energy into high quality protein. Phytoplankton and zooplankton contain high percentages of protein (40-60%) and there is reason to believe that the protein requirements of plankton feeding species are also similarly high. The real difference between species of different feeding habits would appear to be in the ability to digest carbohydrates. Juveniles and adults of most cultured crustaceans have protein requirements in the range of 30-50% of their dry diet weight. Crustaceans appear to have a limited ability to store protein. There are wide variations in the protein requirements of shrimp species.

PLANT PROTEINS

Information regarding the use of leaf proteins in fish nutrition is, yet, negligile except for some vegetable eating species, but because of their high production and competitive economy in agricultural industries, they may in the near future occupy a prominent place in fish feeds after adequate processing involving separation of pigments, flavour and toxins.

Algae constitutes the feed of certain varieties of cultural fish. Chorella species have been found to contain all the essential amino acids and protein of desired nutritional and functional quality can be obtained by selecting the suitable media for their culture and adjusting the harvesting time.

Single cell protein are derived fron yeast, bacteria, fungi or algae grown on a variety of substrata, which include hydrocarbons like crude oil, gas oil, natural gas, coal, carbohydrates.

AMINO ACIDS

Ingested proteins are first split into smaller fragments by pepsin or trypsin or chymotrypsin from the pancreas in the fish. These peptides are then further reduced by the action of carboxypeptidase and aminopeptidase, which hydrolyses of amino acids released into the digestive system are then absorbed through the walls of the gastro-intestinal tract into the blood stream, where they are then resynthesized into new tissue proteins, or catabolised for energy or fragmented for further tissue metabolism. Amino acids are described as the building blocks of proteins and about 23 of them have been isolated from natural proteins. 10 of these amino acids are considered indispensable for fish, crustaceans and molluscs; these are arginine, histidine, isoleucine, leucine, lysine, methionine, phenylalanine, threonine, tryptophan and valine, alanine, aspartic acid, cystine, glutamic acid, glycine, proline, serine and tyrosine are considered to be the non-essential amino acids. Several investigations have shown the potential for supplementing amino acids deficient proteins with limiting amino acids in diets for several fish. Casein supplemented with 6 AA's gave feed conversion ratios similar to isolated fish proteins as a dietary source. It was demonstrated that soyabean meal supplemented with five or more AA's, increasing methionine and lysine.

TABLE .9.1: SOME QUANTITATIVE ESSENTIAL AMINO ACID REQUIREMENTS OF CERTAIN SPECIES OF FISH IN g/KG DIET

AA	*Salmon*	*Eel*	*Carp*
Arginine	24	17	16
Histidine	7	8	8
Isoleucine	9	15	9
Leucine	16	20	13
Lysine	20	20	22
Methionine	16	12	12
Phenylalanine	21	22	25
Threonine	9	15	15
Tryptophan	2	4	3
Valine	13	15	14

TABLE 9.2 : COMPARATIVE AMINO ACID REQUIREMENTS, AS PERCENTAGE OF PROTEIN

AA	*Eel*	*Salmon*	*Catfish*	*Carp*
Arginine	4.2	6.0	4.3	4.3
Histidine	2.0	1.8	1.5	2.1
Isoleucine	3.8	2.3	2.6	2.5
Leucine	4.7	4.0	3.5	3.6
Lysine	5.1	5.0	6.8	5.7
Methionine + Cystine	4.8	3.8	2.9	3.1
Phynylalanine + Tyrosine	5.8	5.3	4.5	6.5
Threonine	3.8	2.3	2.2	3.9
Tryptophan	1.1	0.5	0.5	0.8
Valine	3.8	3.3	3.0	3.6

The absolute AA requirements of crustaceans have yet to be defined. Different AA's are commonly used in proteins in crustacean purified diets.

TABLE. 9.3 : AMINO ACIDS COMMONLY USED IN CRUSTACEAN DIETS, IN PERCENTAGE OF TOTAL WEIGHT OF AMINO ACIDS

AA	*Brine shrimp*	*Casein*	*Egg albumin*	*Shrimp meal*
Aspartic acid	10.1	7.1	10.7	11.3
Threonine	4.9	4.9	4.0	4.2
Serine	5.2	6.3	8.2	4.3
Glutamic acid	14.6	22.4	16.5	16.6
Proline	4.7	10.6	3.8	3.7
Glycine	4.9	2.0	3.6	4.9
Alanine	5.2	3.2	7.6	5.8
Cysteine	1.3	0.3	2.8	1.1
Valine	5.3	7.2	8.8	6.1
Methionine	2.3	2.8	5.3	2.8
Isoleucine	5.1	6.1	7.0	4.5
Leucine	8.6	9.2	9.9	7.8
Tyrosine	4.6	6.3	4.1	4.1
Phynylalanine	5.3	5.0	7.2	4.7
Lysine	7.4	8.2	6.5	8.2
Histidine	2.2	3.1	2.9	2.3
Arginine	6.8	4.1	6.0	7.9
Tryptophan	0.0	1.7	1.2	0.0

LIPIDS AND ESSENTIAL FATTY ACIDS

Lipids are a group of fat soluble compounds occurring in the tissues of plants and animals and broadly consists of fats, phospholipids, sphingomyelins, waxes, esters of glycerol and are the principal form of energy storage. They contain more energy per unit weight than any other biological product—it is estimated that they provide 8.5 kcals metabolisable energy (ME) per gram. Natural diets may contain as much as 50%. Phospholipids are the esters of fatty acids and phosphatic acid. Fatty acids are described as saturated when they contain no double bonds and unsaturated when they contain one (mono-unsaturated) or more polyunsaturated) double bonds. They are composed of carbon, hydrogen and oxygen and are generally acyclic, unbranched molecules containing an even number of carbon atoms. The polyunsaturated fatty acids (PUFA) are divided into 3 families, named after the shortest chain length fatty acid representing each, namely, oleic, linoleic and linolenic acids. The omega (w) system of nomenclature is used to identify the families. Those belonging to oleic family are referred to as w9, linoleic as w6 and linolenic as w3 fatty acids. An abbreviated form is used to refer to the structure of the fatty acid, for example, linoleic acid is expressed as 18:2w6, where 18 is the number of carbon atoms in the fatty acid molecule, 2 is the number of double bonds and w6 is the location of the first double bond relative the methyl (CH_3) end of the molecule. Fish in general contain more w3 than w6 PUFA, but freshwater fish appear to have higher levels of w6 fatty acids than marine species. Most of the estimates of essential fatty acids (EFA) requirements for crustaceans are suggestive rather than conclusive. There are very few reports on the effects of purified fatty acids in semi-purifie test diets for crustaceans. Kanazawa *et al.* (1977) found that both 18:2w6 and 18:3w3 improved the growth of *Penaeus japonicus*. The lipid contents of most commercial aquaculture diets are less than 10% mainly due to processing problems.

CLASSIFICATION

Lipids may be classified into two basic groups according to presence or not of the alcohol glycerol.

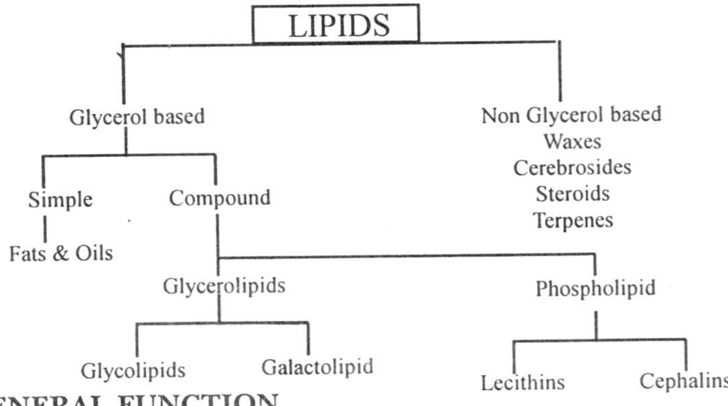

GENERAL FUNCTION

Lipids are important sources of metabolic energy (ATP). In fact, the lipids are the most energy rich of all classes of nutritients : gross energy value of

Lipid	9.5 K Cal/g
Proteins	5.6 K Cal/g
Carbohydrate	4.1 K Cal/g

In this respect, dietary lipids may be used to spare the more valuable protein for growth. In particular, free fatty acids derived from triglycerides (fats and oils) are the major aerobic fuel source for energy metabolism of fish muscle. Lipids are essential components of all cellular and sub cellular membranes (lipid classes that are involved include the polyunsaturated fatty acid containing phospholipids and sterol esters). Lipids serve as biological carriers for the absorption of the fat soluble vitamins A, D, E and K. Lipids are a source essential fatty acids, which in turn are essential for the maintenance and integrity of cellular membranes are required for optimal lipid transport (bound to phospholipids as emulsifying agents) and are precursors of the prostaglandin hormones.

Lipids are believed to play a role as a mechanical cushion or support for the vital body organs, and aid in the maintenance of neutral buoyancy.

Lipids are a source of essential steroids which in turn perform a wide range of important biological functions (that is, the sterol cholesterol is involved in the maintenance of membrane systems, for lipid transport and as a precursor of vitamin D_3, the bile acids and the steroid hormones (androgens, estrogens, adrenal hormones and corticosteroids). From feed technology view point, lipids act as a lubricants for the passage of feed through pellet diets, as substances which reduce the dustiness of feeds and paly a role in feed palatibility.

FATS AND OILS

Fuel or energy can be stored in plants as starch, and in animals as glycogen, but it can also be stored in both plants and animals in a more compact form as fats or oils. In plants, fats and oils are farmed from carbohydrate (Eg: As plants seeds ripen, their starch content falls as their fat content increases). In animals fats can be formed also from carbohydrates (i.e., fattening a pig with food largely composed of carbohydrates). However, unlike plants, animals can also deposit fat in their body from fat ingested. The only difference between fats and oils is that the latter are liquid at room temperature whereas fats are semisolid at room temparature.

COMPOSITION

Fats and oils normally occur in food stuffs and in the fat deposits of most animals in the form of triglycerides, which are esters of fatty acids and glycerol.

$$\begin{array}{lclcl} CH_2OH & & R\text{—}COOH & & CH_2\ .\ O\ .\ COR & & \\ | & & & \longrightarrow & | & & \\ CHOH & + & R\text{—}COOH & & CH_2\ .\ O\ .\ COR & + & 3H_2O \\ | & & & & | & & \\ CH_2OH & & R\text{—}COOH & & CH_2\ .\ O\ .\ COR & & \\ \text{Glycerol} & & \text{3 Fatty Acids} & & \text{Triglyceride (Fat \& Oil)} & & \text{Water} \end{array}$$

Naturally occurring fats and oils are composed of mixed triglycerides where glycerol is esterified with different types of fatty acids, i.e., fatty acids R_1, R_2 and R_3 thus:

$$
\begin{array}{l}
CH_2 . O . COR_1 \\
| \\
CH . O . COR_2 \\
| \\
CH_2 . O . COR_3
\end{array}
$$

Mixed Triglyceride

Naturally occurring fat or oil found in nature consists of single triglycerides. It can be seen that the basic unit and variable of all triglycerides is the fatty acid component, which in turn affect the physical and chemical property of the fat or oil.

FATTY ACID STRUCTURE AND CLASSIFICATION

Over 40 different fatty acids are known to occur in nature, they all can be represented by general formula.

$$CH_3\ (CH_3)_n\ COOH$$

Where n = 0 in Acetic acid

n = 1 in Propionic acid

n = 2 in Butyric acid, etc.

Where n is usually an even number. Most naturally occurring fatty acids contain a single - COOH group and a straight unbranched carbon chain (C), which may in turn contain no double bond (saturated fatty acid), a single double bond (mono-unsaturated fatty acid) or more than one double bond (poly unsaturated fatty acid, PUFA). It follows that the degree of unsaturation will greatly influence the physical properties of the constituent fats as in general unsaturated fatty acids are more chemically reactive and have lower melting points than the corresponding saturated fatty acids. Examples of saturated and unsaturated fatty acids are given below

Fatty acid	*Structure*	*Shorthand abbreviation*[1]
Saturated		
Butyric acid	$CH_3(CH_2)_2COOH$	4:0
Caproic acid	$CH_3(CH_2)_4COOH$	6:0
Capric acid	$CH_3(CH_2)_8COOH$	10:0
Lauric acid	$CH_3(CH_2)_{10}COOH$	12:0

{Cont.}.....

Myristic acid	$CH_3(CH_2)_{12}COOH$	14:0
Palmitic acid	$CH_3(CH_2)_{14}COOH$	16:0
Staeric acid	$CH_3(CH_2)_{16}COOH$	18:0
Unsaturated		
Palmitoleic acid	$CH_3(CH_2)_5CH=CH\ (CH_2)_7\ COOH$	16:1 n-7
Oleic acid	$CH_3(CH_2)_7\ CH=CH\ (CH_2)_7\ COOH$	18:1 n-9
Linoleic acid	$CH_3(CH_2)_4\ CH=CH\ (CH_2)_7\ COOH$	18:2 n-9
Linolenic acid	$CH_3\ CH_2\ CH=CH\ CH_2\ CH=CH\ CH_2$ $CH=CH\ (CH_2)_7\ COOH$	18:3 n-3
Arachidonic acid	$CH_3(CH_2)_4\ CH=CH\ CH_2\ CH = CH\ CH_2$ $CH=CH\ CH_2\ CH = CH\ (CH_2)_3\ COOH$	20:4 n-6
Eicosapentanoic acid	$CH_3\ CH_2\ CH=CH\ CH_2\ CH=CH\ CH_2\ CH=CH\ CH_2$ $CH=CH\ CH_2\ CH=CH\ (CH_2)_5\ COOH$	20:5 n-3
Docosahexaenoic acid	$CH_3\ CH_2\ CH=CH\ CH_2\ CH =CH\ CH_2\ CH = CH\ CH_2$ $CH=CH\ CH_2\ CH= CH\ CH_2\ CH=CH\ (CH_2)_2\ COOH$	22:6 n-3

1. Number of carbon (C) atoms: number of double bonds and position of the first double bonds and position of the first double bond counting from the methyl (CH_3) end of the fatty acid.

On the basis of the above classification polyunsaturated fatty acids (PUFA) may be divided into three major families; the oleic (n-9) series, the linoleic (n-6) series and the linolenic (n-3) series, the family names representing the shortest chain number of the group, with other family members being derived from these three basic groups.

FATTY ACID BIOSYNTHESIS

With the exception of the land snail *(Cepaea nemoralis)* animals are incapable of *de novo* synthesis of fatty acids with double bonds in the n-6 (linoleic series) and n-3 (linolenic series) position; only plants are able to synthesize these fatty acids *de novo*. However, most animals are able to synthesize even chain saturated fatty acids from acetate, or of adding two carbon units to the carboxyl end of a fatty acid and adding more double bonds to the carboxyl side of the existing double bonds but not on the methyl end. The biochemical pathways of PUFA biosynthesis in fish and prawn can be summarised as follows.

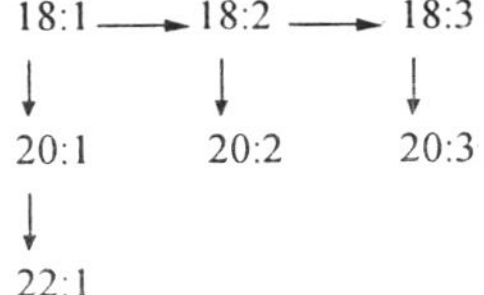

Oleic acid n-9

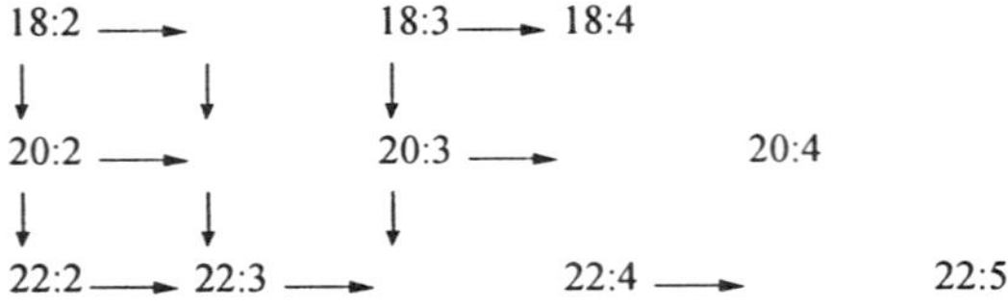

Linoleic acid n-6

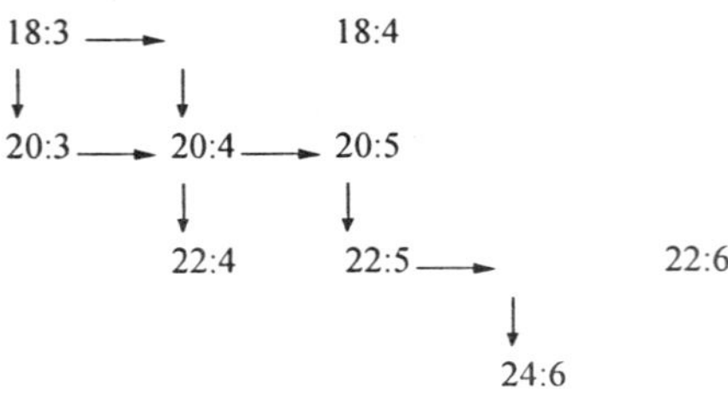

Linolenic acid n-3

***Vertical arrows show chain elongation reactions. Horizontal arrows show desaturation reactions.**

ESSENTIAL FATTY ACID REQUIREMENT

In view of inability of animals to synthesize *de novo* fatty acids of the n-6 and n-3 series, these fatty acids must be supplied in a ready made form within the diet. For land animals, the linoleic (n-6) series has been found to have the highest essential fatty acid (EFA) activity, with the linolenic (n-3) series having only partial EFA activity. It follows, therefore, that the predominant fatty acids (PUFA) in the tissues of land animals being to the linoleic series, namely, 18:2 n-6 (linoleic acid) and 20:4 n-6 (Arachidonic acid).

By contrast, the predominant PUFA in the tissues of fish and prawn belong to the linolenic (n-3) series, and this applies to freshwater and marine fish alike. The concentration of n-6 PUFA in the tissue of fish is generally low, although higher levels are reported in freshwater fish species. This is perhaps not surprising if one considers that the diet of freshwater fish contains a component derived from terrestrial sources, and consequently rich in n-6 series of fatty acids. It is generally believed that the n-3 series fatty acids permit a greater degree of unsaturation a requirement for greater membrane fluidity, flexibility and permeability at low temperatures. In fact it is generally believed that the dietary requirement (preferential) of fish for n-3 series EFA, over n-6 series is fundamentally due to the low water temperature of their aquatic environment (as compared with mammals). In fact, the lower the water temperature, the greater the incorporation of n-3 series PUFA in the tissues. Apart from the differences in n-6 PUFA content of the tissues of freshwater and marine fish species, freshwater fish also generally have higher tissue concentration of the shorter chain PUFA n-3 series.

TABLE 9.4: DIETARY ESSENTIAL FATTY ACID (EFA) REQUIREMENT OF FISH

Fish		*Requirement*						
Rainbow Trout		1%	18:3	n–3				
	or	1%	HUFA	n–3				
Ayce		1%	18:3	n–3				
	or	1%	20:5	n–3				
Common carp		1%	18:3	n–3	+	1%	18:3	n–3
	or	0.5	—	1%	HUFA	n–3		
Eel		0.5%	18:2	n–6	+	0.5%	18:3	n–3

With the exception of strict carnivorous fish species, fish are able to chain elongate and further dasaturate 18:2 n-6 or 18:3 n-3 (depending on the fish species) to the corresponding highly unsaturated fatty acids (HUFA): 20:4 n-6 in the case of the n-6 series and 20:5 n-3 or 22:6 n-3 in the case of the n-3 series. It is generally believed that these HUFA are responsible for the key metabolic functions ascribed to the EFA. In fact, for most fish species HUFA have greater EFA activity than the corresponding basic unit (18:2 n-6 or 18:3 n-3).

In general, cold water freshwater fish have an exclusive requirement for n-3 series PUFA (18:3 n-3, 20:5 n-3, 22:6 n-3) in their diet (i.e., Salmonoids, Ayu) while warm freshwater fish have either a requirement for both the n-3 series and n-6 series alone (i.e., Tilapias, and possibly the snake head, *Channa micropeltes)*, in the case of marine carnivorous fish species (i.e., Red seabream, black seabream *Mylio macrocephalus*, opaleye *Girella nigricans*, puffer fish, *Fugu rubripens*, yellow tail *Seriola quinqueradiata*, plaice *Pluronectes*. Plastessa, gilthead bream *Sparus auratus*, turbot *Scophthalmus maximus*). Since the food organisms consumed are rich in 22:6 n-3 and 20:5 n-3 they have lost ability to chain elongate and further desaturate 18:3 n-3 to the corrresponding HUFA. Marine carnivorous fish must, therefore, be supplied with 22:6 n-3 or 22:5 n-3 in a ready made form.

PHOSPHOLIPIDS

Within the animal body, the phospholipids represent the second largest lipid component after the triglyceride fats and oils. All phospholipids are yellow greasy solids and share the property of being soluble in lipid solvents with the exception of acetone (this property allows them to be readily distinguished from fatty acids).

Like fats and oils, phospholipids are all esters of fatty acids and glycerol. However, whereas in simple fats and oils the trihydric alcohol glycerol is esterified with three fatty acids, in phospholipids only two of the alcohol groups of glycerol are esterified with fatty acids; the remaining group being esterified with phosphoric acid and a nitrogenous base. According to the nitrogenous base present, phospholipids may be divided into two groups; lecithins (nitrogenous base—ethanolamine). Their structures can be represented as:

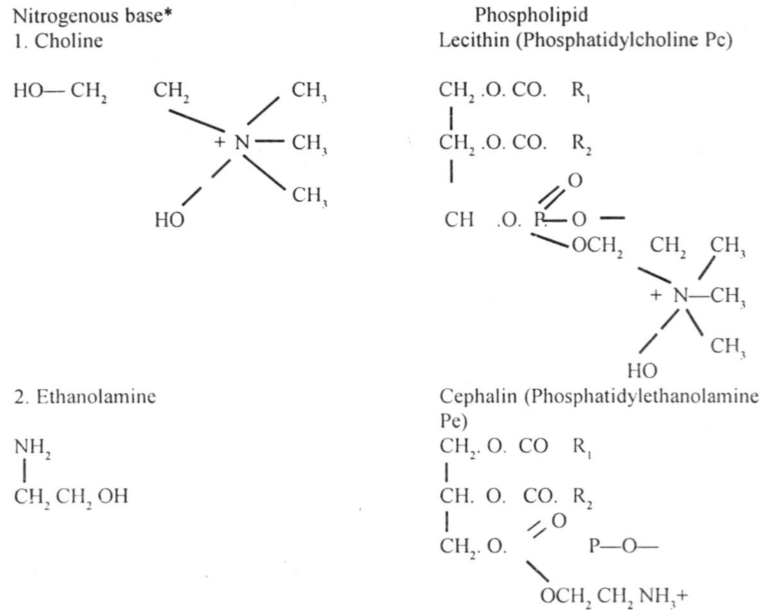

Other nitrogenous bases may include serine and inositol.

From these structural formulae, it can be seen that the phospholipids like fatty acids, have a polar region and a non-polar region. However, unlike the fatty acids, the ionic functions are greatly increased by the presence of phosphoric acid and the nitrogenous organic base; which therefore results in combining within the same molecule, both hydrophilic and hydrophobic (fatty acid chain) sites. It is because of this unique surface active property that phospholipids in conjugation with proteins, from the basic lipoprotein structure of biological membranes. It is interesting to note here that the fatty acids contained within animal phospholipids (R_1, R_2) are much more unsaturated than corresponding fatty acids from triglycerides (fats and oils). The increased unsaturation of the phospholipids acids is largely due to the increased levels of the C_{20} and C_{22} polyunsaturated fatty acids, which are almost exclusively bound at the 2 position. In particular, the fatty acids 20:5 n-3 and 22:6 n-3 (EFA) may account for 80% of the total fatty acid found at the 2 position. It follows, therefore that during EFA deficiency, examination of the phospholipids in tissues shows the presence of high levels of polyunsaturated fatty acids derived from oleic and palmitoleic acid, in contrast to the usual situation where polyunsaturated fatty acids derived from linolenic acid predominate. Phospholipids also play important roles as emulsifying agents in biological systems and are particularly involved in the transportation of fats within the body. For example phospholipids may take part in the emulsification of dietary lipids in the digestive tract, and as constituents of high density lipoproteins in the transport of lipids within the body. Rich dietary sources of phospholipids include eggs and soybean oil.

DIETARY REQUIREMENT

Dietary phospholipids have been found to have a beneficial effect on the growth and survival of marine fish larvae (red sea bream) and marine shrimp (*P. japonicus*). *P. monodon* fed semi-synthetic test diets in which choline and the EFA were added separetly at equivalent levels. Furthermore, the efficacy of phospholipids on growth and survival has been shown to vary with the type and source of phospholipids used. For example, the efficacy of bonito-egg PC, soybean PC and soybean phosphotidylionsitol (PI) has been found to be much higher than that of bonito-egg PE, ovine-brain PE, ovine-brain phosphotidylserine (PS) or chicken egg PC in *P. japonicus* larvae. These researchers also showed that the optimum level of dietary phospholipid for *P.japonicus* larvae varied with the dietary lipid source used; thus a dietary soyabean PC requirement of 6.0% and 3.5% was obtained when 18:1 n-9 and 1.0% highly unsaturated fatty acids, or pollack liver oil was used as the dietary lipid source, respectively A 3% dietary requirement for soybean Pc has also been reported for Aya and *P. japonicus* larvae when pollack liver oil was used as the basal lipid source. Kanazawa (1985) concluded that (1) phospholipids containing either choline or inositol exerted a positive effect on growth and survival, (2) phospholipids containing 18:2 n-6, 18:3 n-3, 20:5 n-3 and 22:6 n-3 in the molecule were the most effective in promoting growth and survival and (3) the effectiveness of the phospholipids seemed to be dependent on the nature of the fatty acids in the and position of the phospholipid molecule. The beneficial effect of dietary phospholipids on the growth and survival of marine fish larvae and crustaceans is particularly surprising bearing in mind the natural capacity of these animals for phospholipids biosynthesis from fatty acids and diglycerides. Although a true requirement for dietary phospholipids remains to be confirmed under practical farming conditions, it has been suggested that the dietary essentiality of phospholipids (if at all) is due to a specific requirement for phospholipids for fatty acid transport within the body and a slow rate of biosynthesis of phospholipids in relation to metabolic demand during the larval growth phase.

Glycolipids

Glycolipids are similar to phospholipids in that they are glycerol based and have two of the alcohol groups present esterified by fatty acids, but differ from them in that the third group is linked to a sugar residue.

CH_2.O. CO .R
|
CH.O.CO. R
|
CH_2O—

CH_2OH
O
OH
H
OH
H
H
OH
H

Galctolipid

The lipids of grasses and clovers, which form the major part of the dietary fat of ruminants, are predominantly (60%) galactolipids. In general, about 95% of the fatty acid present in linoleic acid (18:2n-6).

Waxes

The waxes are esters of fatty acids with high molecular monohydric alcohols. Like fats, natural waxes occur as mixtures of different esters, which are usually solid at room temperature. Waxes are widely distributed in both plants and animals, where they generally perform protective roles. For ex. waxes often occur within the cuticle of leaves and fruit so as to minimise water losses due to transpiration; whereas in animals, wool and feathers are often protected against water by the hydrophobic nature of wax coating. Among the better known animal waxes are lanolin (obtained from wool) beeswax (an insect secretion) and spermaceti from the sperm whale. In some aquatic animals, waxes often replace the triglycerides to some extent. For example, in some whales and many crustaceans such as the copepod *Calanus* sps wax esters form the major component of the depot fat. Although waxes are not readily hydrolysed by terrestrial animals and so have no real nutritive value, certain aquatic animals such as marine fish (i.e., sardines, herring, salmon) which prey upon wax rich animals (i.e., copepods) do possess wax esterns lipases which are able to cleave the wax esters and so make them available for digestion. However, since the fatty acid component of these wax esters is generally saturate, and so deficient in the long chain PUFA, they proably only serve as energy source, rather than for structural purposes.

Steroids

The steroids include a very important and widely distributed group of substances including the sterols, the bile acids, the adrenal hormones and the sex hormones. Although these steroids have a very wide range of biological properties they all have basic structural unit of a phenanthrene nucleus linked to a cyclopentane ring.

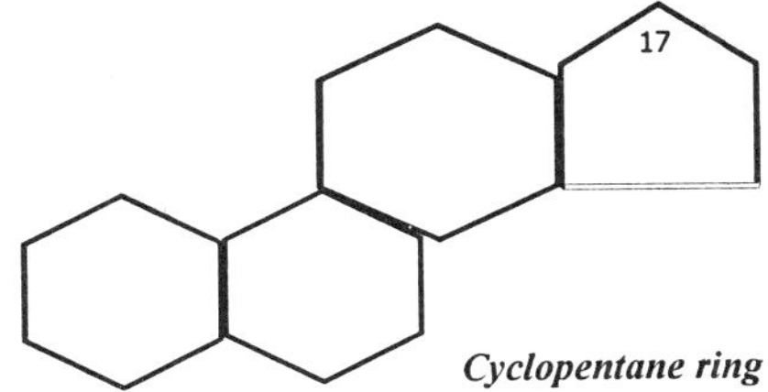

Cyclopentane ring

The individual compounds differ in the number and position of their double bonds and in the nature of the side chain at carbon atom 17.

Cholesterol

Cholesterol is widely distributed in the animal body, being particularly abundant in the brain and nervous tissue, blood, bile, liver, and the skin. Within the body cholesterol may exist in its free state (i.e., cholesterol is the chief component of gall stones) or in esterfied form fatty acids and other organic acids. Cholesterol performs many important functions within the body. It is an essential component of biomembranes system in all eukaryotic species, together with the phospholipids and proteins. The bulk of the cholesterol in animal tissues is associated with membrane systems. Many important sterols found within the body are synthesised from cholesterol. For example, cholesterol is a precursor of the bile acids, steroid hormones (including the androgens, estrogens, and corticosteroids) and of vitamin D_3.

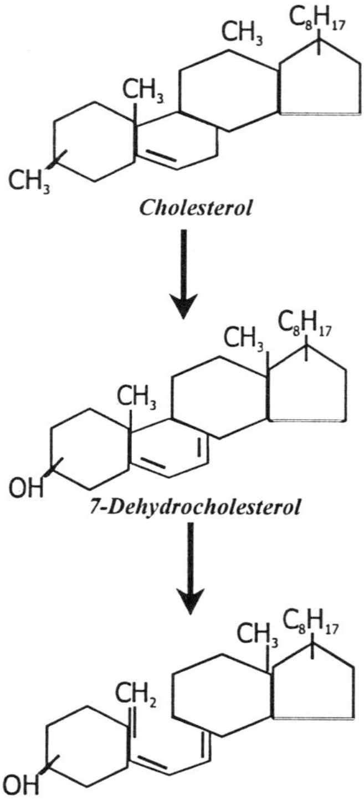

CH_3 C_8H_{17} CH_3 CH_3

Cholesterol

↓

CH_3 C_8H_{17} CH_3 OH

7-Dehydrocholesterol

↓

CH_3 C_8H_{17} CH_2 OH

Cholecalciterol (Vitamin D_3)

Cholesterol also plays an important role in the absorption of fatty acids from, the intestine and in their consequent transportation in the blood or haemolymph. Here cholestol combines with fatty acids to form cholesterol esters, which are more soluble and emulsifiable than the free fatty acid molecules. In contrast to fish, crustaceans like other arthropods have been found to be incapable of synthesizing sterols *de novo* from acetate and mevalonate. Cholesterol is therefore regarded as an essential dietary nutrient for marine prawn and freshwater prawns. Based on laboratory studies conducted with *P. japonicus* the optimum level of dietary cholesterol is reported to be 0.5 - 2% of the dry diet. A rich source of dietary cholesterol is prawn head oil.

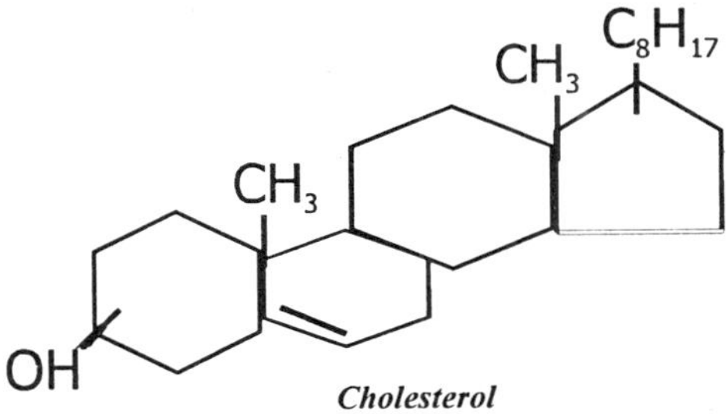

Cholesterol

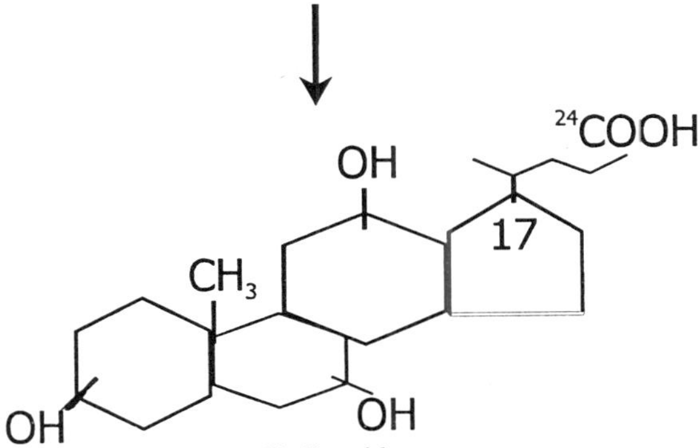

Cholic acid

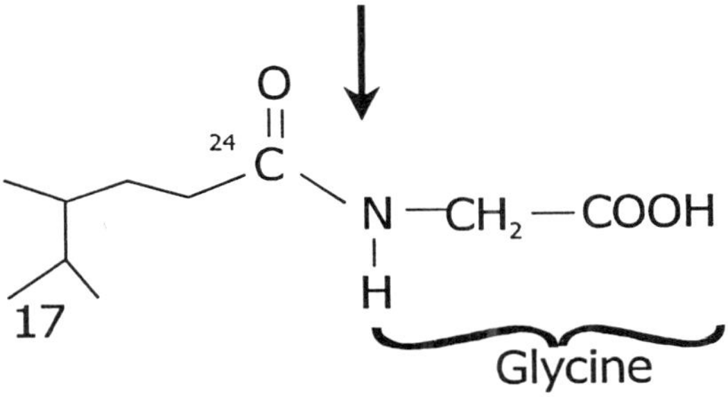

Glycine

Glycocholic acid

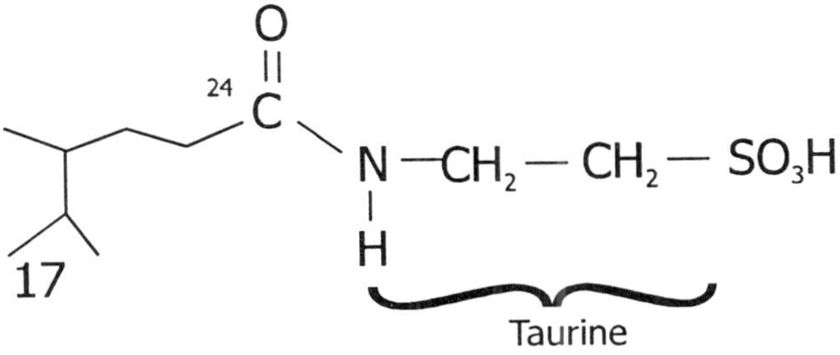

Taurine

Taurocholic acid

Bile acids

These steroids are formed by the combination of the amino acids glycine or taurine with cholic acid (a derivative of cholesterol). The bile acids are formed and concentrated by the liver and are excreted by the liver into the bile and pass to the gastro intestinal tract (duodenum) via the bile duct, where they act as important biological emulsifiers. They help to solubilise fat globules from the food so that the water soluble enzymes or lipases can react with the fat molecules and split them to facilitate fat absorption. Bile acids also facilitate the major excretion route of cholesterol.

Lipid pathology

Dietary essential fatty acid deficiency

All fish and shrimp examined to date display reduced growth and survival, and poor food conversion efficiency when fed experimental diets deficient in the essential fatty acid (EFA). The following additional gross anatomical deficiency signs have been reported under laboratory conditions with juvenile fish and shrimp fed EFA deficient diets.

Fish/prawn	*species*	*EFA deficiency signs*
Rainbow trout	(*S. gairdneri*)	Increased mortality, increased suceptibility to caudal fin erosin by. *Flexebacterium* sps fainting or
shock		syndrome, decreased haemoglobin and red blood cell volume, fatty infiltration/degeneration of liver swollen pale liver, reduced spawning efficiency
Coho salmon	(*O. kisutch*)	Swollen pale liver, increased hepatosomatic index (fatty liver), high mortality
Chum salmon	(*O. keta*)	Swollen pale liver increased hepatosomatic index (fatty liver) high mortality.
Common carp	(*C.carpio*)	Increased mortality, fatty liver
Eel (*A. japonica*)		Increased mortality
Tilapia sps.		Swollen pale liver, fatty liver
Red sea bream		Reduced spawning efficiency (decreased hatching rate/survival)
Turbot		Increased mortality, reduced growth, degeneration of gill epithilium

Dietary EFA deficiency sensality results from poor feed formulation.

Dietary essential fatty acid toxicity

Under laboratory conditions it has been shown that a dietary excess of EFA may exert a negative effect on growth and feed efficiency (rainbow trout).

Toxic non-essential fatty acids

Cyclopropinoic acid is a toxic effect fatty acid found in the lipid of cotton seed products. Experimentally, cyclopropinoic acid has been shown to reduce growth rate in rainbow trout and to act as a potent synergist for the carcinogeniety of aflatoxins. Other pathologists observed with trout include extreme liver damage (pale in colour) with increased glycogen deposition and decreased protein content and a decrease in activity of several key enzymes.

Oxidation of dietary lipids

In the absence of suitable antioxidant protection lipids rich in PUFA are highly prone to auto-oxidation on exposure to atmospheric oxygen. Under these conditions the nutritional benefit of EFA in fact becomes deleterious to the health of the fish or prawn. Feed stuffs rich in PUFA which are particularly susceptible to lipid oxidative damage (oxidative rancidity) include fish oils, fish meal, rice bran and expeller oil seed cakes containing little or no natural antioxidant activity. During the process of lipid auto-oxidation chemical degradation products are formed, including free radicals, peroxides, hydroperoxides, aldehydes and ketones, which in turn react with other dietary ingredients (vitamins, proteins and other lipids) reducing their biological value and availability during digestion. At present oxidative rancidity is believed to be one of the major deteriorative changes which occurs in stored feedstuffs. Numerous gross anatomical, pathological signs have been reported in fish fed rations contains oxidized fish/plant oils with no antioxidant (vit-E) protection.

VITAMINS

Definition and Classification

Vitamins are a heterogenous group of organic compounds essential for the growth and maintenance of animal life. The majority of vitamins are not synthesized by the animal body or at a rate sufficient to meet the animals needs. They are distinct from the major food nutrients (proteins, lipids and carbohydrates) in that they are not chemically related to one another, are present in very small quantities within animal and plant foodstuffs, and are required by the animal body in trace amounts. Approximately 15 vitamins have been isolated from biological materials; their essentiality depending on the animal species, the growth rate of animal, feed composition, and the bacterial synthesizing capacity of the gastro-intestinal tract of the animal. In general, all animals display distinct morphological and physiological deficiency signs when individual vitamins are absent from the diet.

Vitamins may be classified into two broad groups, depending on their solubility; the water soluble vitamins and fat-soluble vitamins.

Water soluble vitamins	*Fat soluble vitamins*
Thiamine (Vitamin B_1)	Retinol (Vitamin A)
Riboflavin (Vitamin B_2)	Cholecalciferol (Vitamin D_3)
Pyridoxine (Vitamin B_6)	Tocopherol (Vitamin E)
Pantothenic acid	Phylloquinone (Vitamin K)
Nicotinic acid (Niacin)	
Biotin	
Folic acid	
Cyanocobalamin (Vitamin B_{12})	
Ionositol	
Choline	
Ascorbic acid (Vitamin C)	

As their name suggests, the fat soluble vitamins are absorbed from the gastro-intestinal tract in the presence of fat and can be stored within the fat reserves of the body whenever dietary intake exceeds metabolic demands; storage increasing with dietary intake to the extent that a toxic condition (Hypervitaminosis) may be produced. By contrast, the water soluble vitamins are not stored in appreciable quantities in the animal body; body stores being rapidy depleted in the absence of regular dietary water soluble sources. Water soluble vitamins toxicities are therefore unlikely.

WATER SOLUBLE VITAMINS

The water soluble vitamins include eight well-recognised members of the vitamin B complex, the water-soluble essential factors choline, ionositol and ascorbic acid and less-defined vitamin activities for fish of p-aminobenzoic acid, lipoic acid, citrin and undefined growth factors. The first eight are required in small amounts in the diet but play important roles in growth, physiology and metabolism.

THIAMIN

Thiamin was isolated from rice polishings. Thiamin hydrochloride is a water-soluble, colourless, monoclinic, crystalline compound with emperical formula $C_{12}H_{18}ON_4SCl_2$. Thiamine is found in many grains and seeds, being most concentrated in the seed coat. Discovery of this growth factor began when persons consuming polished rice developed beriberi; but recovered when a factor in the seed coat was added to the diet. Thiamine is composed of a pyrimidine ring and a thiazole ring. It is synthesized by higher plants, but not by animals. It is relatively sensitive to heat and moisture at pH above 6, so the commercial vitamin is sold in the hydrochloride form to ensure stability.

The active form of the vitamin is thiamine pyrophosphate. Phosphorylation occurs in the liver. Thiamine pyrophosphate acts as a coenzyme for several metabolic decarboxylation and transketolation reactions. It is involved in decarboxylation of pyruvic acid and-ketoglutaric acid in aerobic glycolysis. It also acts as a coenzyme in transketolation in metabolism of glucose through the pentose phosphate shunt. Transketolase activity in erythocytes and kidney is a sensitive indicator for thiamine status in rainbow trout. Thiamin deficiency affects the central nervous system in fish, birds and mammals. Polyneuritis in chicks (lack of control of position of the head) and chastek paralysis in mink and foxes are caused by thiamin deficiency. Thiamin deficiency in fish causes hypersensitivity to disturbance, loss of equilibrium, and convulsions. Fish show thiamin deficiency quickly; channel catfish in 6 to 8 weeks, carp in 8 weeks, and Japanese eel in 10 weeks.

Tissues of most fish contain thiaminase, an enzyme that can destroy thiamin in nonliving tissue by splitting it into its two component ring structures. Heating the fish destroys the enzyme. Feeding fish or fish visceral organs without heat treatment has produced thiamin deficiency problems in mink and foxcs and in channel catfish. The thiamin is destroyed prior to ingestion, when the enzyme (thiaminase) and substrate (thiamin) are in contact for a period of time. Channel catfish fed diets containing 40% non heated fish viscera developed thiamin deficiency in 10 weeks, but when the fish were fed an additional diet containing thiamin in a desperate meal daily, no deficiency occurred.

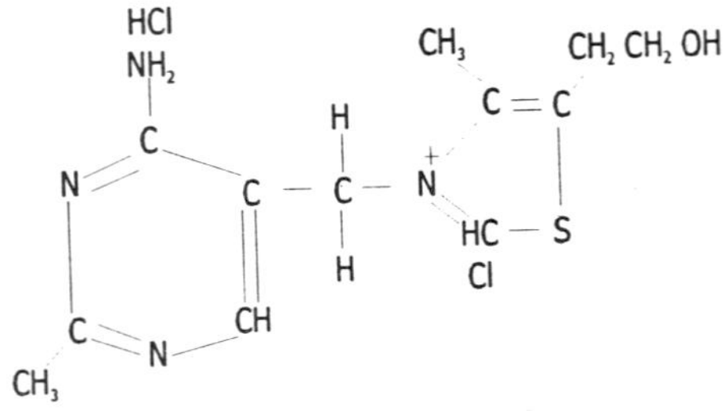

Thiamin hydrochloride

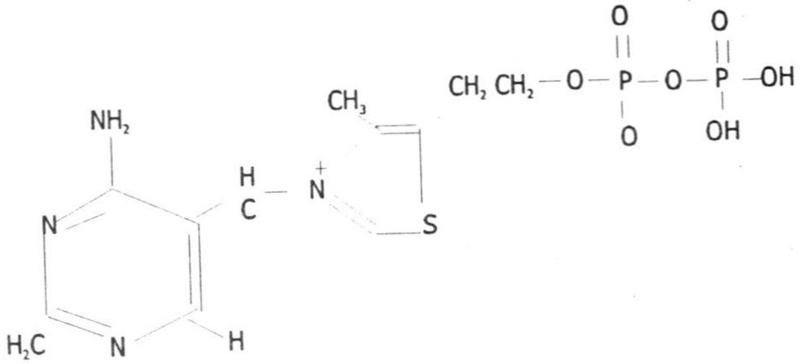

Carboxylase or Thiamin pyrophosphate

RIBOFLAVIN

Riboflavin is a yellow-brown crystalline pigment with chemical formula $C_{17}H_{20}N_4O_6$. Riboflavin is widely found in nature: green plants, seed coat of grains, and yeast are rich sources. Its origin is plant or micro organism synthesis. Non ruminant animals require it in their diet. Chemical structure of riboflavin consists of a dimethyl-isoalloxazine moiety conjugated with ribose through which the vitamin is linked to phosphate in the intestinal wall to form the active coenzyme.

Riboflavin is a component of two flavoprotein coenzymes, flavin mononucleotide (FMN) and flavine adenine dinucleotide (FAD), which are components of prosthetic groups of oxidases and reductases that act upon metabolic degradation products of proteins, carbohydrates and lipids.

A number of riboflavin deficiency signs have been identified in animals, but none have related to a specific biochemical role of the vitamin. Crooked and stiff legs occur in swine and chickens (curled-toe paralysis). Eye problems occur in humans, farm animals, and fish photophobia and cataracts have been found in several fish deprived of riboflavin. Dark skin occurred in salmonids; stubby bodies were described in riboflavin deprived channel catfish; dermatitis and hemorrhagic fins were found in eel; and necrosis of head kidney was observed in carp.

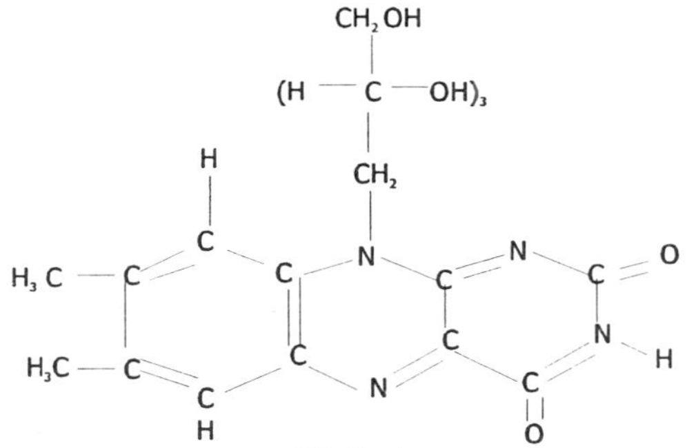

Riboflavin

A sensitive subclinical test for riboflavin deficiency is *in vitro* measurement of erythrocyte glutathione reductase (EGR) activity in presence of added FAD. Approximately 3 mg of riboflavin per kg of diet is sufficient of to prevent change in EGR activity in rainbow. The mininum dietary requirement for normal growth and to prevent dwarfism in channel catfish is a mg/kg of body weight. Most feed stuffs are reasonably good sources of riboflavin, except whole grains. Bran or polishings from grains and distillery byproducts are fair to good sources.

NIACIN

Niacin is pyridine-3-carboxylic acid with chemical formula $C_6H_5O_2$. Niacin, nicotinic acid and nicotinamide are often used interchangeably. Niacin is actually the generic descriptor for pyridine 3-carboxylic acid and derivatives exhibiting biological activity of nicotinamide. Nicotinamide occurs in physiological systems as a component of two coenzymes of the hydrogen transport system. Nicotinamide adenine dinucleotide (NAD) and Nicotinamide adenine dinucleotide phosphate (NADP). These coenzymes are involved in a number of oxidation reduction reactions. NAD is specific for hydrogenases involved in passing electrons on to oxygen in the electron transport systems. NADP is specific in hydrogenases such as in fatty acid synthesis and the pentose phosphate shunt in glucose metabolism. Niacin deficiency causes pellagra in humans which is characterized primarily by dermatological problems, and is commonly found in situations where the diet is limited to corn or polished rice. Most fish show niacin deficiency signs rather quickly. Skin lesions are a common deficiency sign in fish. Channel cat fish show skin and fin lesions along with deformed jaws, exophthalmia, anemia, and high mortality rate. Eels show skin lesions, dark pigmentation and ataxica, salmonids show sensitivity of skin to sunburn along with fin erosion, intestinal lesions and muscle weakness.

N
HC CH
HC C —COOH
C
H

Nicotine acid

N
HC CH
C — $CONH_2$
HC
C
H

Niacinamide

Niacin

Niacin occurs naturally as an amide in all living tissue. Although plant seeds contain substantial amounts of niacin, the niacin occurs mostly in bound form and is highly unavailable to animals, most land animals can convert the amino acid tryptophan to niacin and this source can contribute to their niacin requirement. However, fish appear to have limited ability to make this conversion. Brook trout were found incapable of converting tryptophan to niacin efficiently. Also niacin deficiency can be demonstrated rapidly in fish fed tryptophan in the diet. Because of the presumed limited availability of niacin in grains and oil seed meals to fish, bits supplementation of practical fish feeds is recommended.

PANTOTHENIC ACID

Pantothenic acid is synthesized by plants and microorganisms chemically, it is composed of pantoic acid [2, 4-dihydroxy - 3,3-dimethyl (butyric acid)] linked through a peptide bond to Blanine. In nature, it occurs largely in bound form as coenzyme A. Because of its sensitivity to destruction by heat and high or low pH, it is available commercially as calcium or sodium salt.

The only known function of pantothenic acid is as a component of coenzyme A (CoA), which is involved in transfer of acetyl (2-carbon) units in numerous reactions in the metabolism of proteins, carbohydrates and lipids. Coenzyme A contains pantothenic acid linked through pyrophosphate to adenosine 3'-phosphate on one side andβ-mercaptomethylamine on the other. The acceptance of the acetyl units, which help form acetyl CoA, is through the sulfohydryl group of β-mercaptomethylamine. Coenzyme A formation is an important route for carbohydrates and fatty acids to enter the tricarboxylic acid cycle. Coenzyme A has manifold functions, anabolic and catabolic, involving cellular energy release and synthesis of steroids, sphingosine, porphyrin and many other compounds.

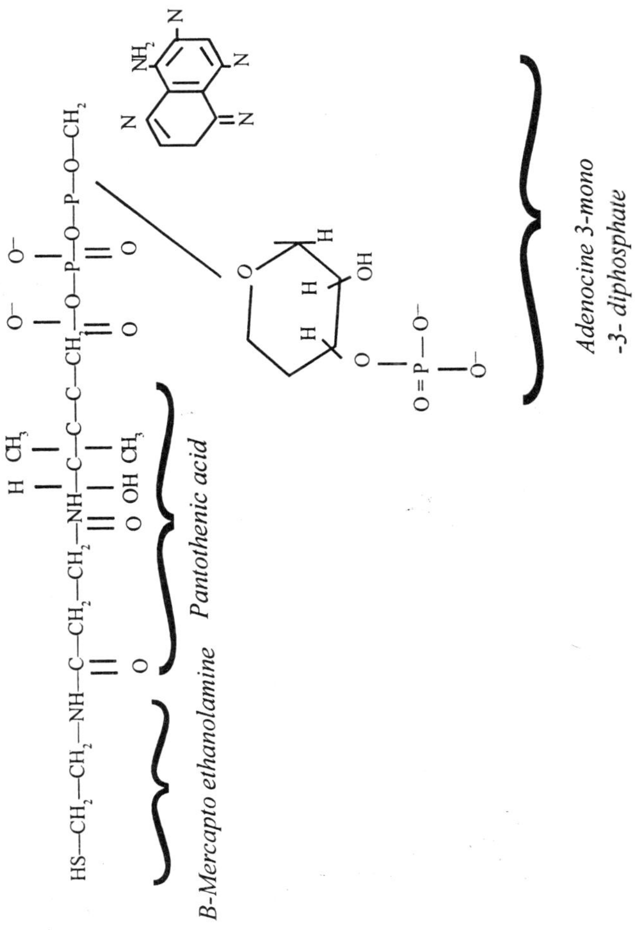

$HS—CH_2—CH_2—NH—C—CH_2—CH_2—NH—C—C—C—CH_2—O—P—O—P—O—CH_2$

B-Mercapto ethanolamine

Pantothenic acid

Adenocine 3-mono -3- diphosphate

Pantothenic acid is a dietary component essential for birds, non-ruminant mammals and fish. A dietary deficiency results in growth failure in all skin and hair problems are common deficiency characteristics. Pantothenic acid deficiency results in impaired functions of mitochondrial-rich, high energy expenditure cells. This may partially explain why gill lamellae, kidney tubules, and acinar cells in pancrease, all of which carry on a high level of metabolic activity, are all sensitive to pantothenic acid deficiency. Pantothenic acid is widely found in commercial feed stuffs, but the level and availability in processed feed are likely to be lower than the requirement for most fish. Supplemental pantothenic acid is recommended for commercial fish feeds, especially for small fish.

B_6 (PYRIDOXINE)

Three chemically related compounds that have similar metabolic functions have been identified: pyridoxine, pyridoxamine and pyridoxal. Pyridoxine was the first identified and was given this name. Later, the other two were identified and the name B_6 was given to this group of compounds. Vitamin B_6 is now the approved name for this vitamin. Vitamin B_6 is a dietary requirement of all non-ruminant mammals, birds and fish. Vitamin B_6 compounds are synthesized by plants and some microorganisms. It is widely found in feed stuffs of plant and animal origin. The metabolically active B_6 coenzyme is pyridoxal phosphate. It is functional in a number of enzymes in which amino acids are metabolized, including decarboxylases, transaminases, sulfhydrases and hydroxylases. Increasing protein percentage in the diet causes an increase in the B_6 requirement of salmonids. Because fish are fed much higher protein diets than land animals, it should be expected that the B6 requirement would be higher for fish. Vitamin B_6 is involved in metabolism of carbohydrates and lipids. It is essential for the synthesis of heme (in hemoglobin), and in the synthesis of serotonin from tryptophan, which may explain why B_6 deficiency causes nervous disorders.

Deficiency signs in fish develop quickly young channel cat fish, salmonids, and carp show deficiency signs in 4 to 8 weeks. These include nervous disorders, such as hypersensitivity to disturbances: poor swimming co-ordination: convulsions; and tetany when handled. Channel cat fish develop a greenish-blue sheen to the skin. Common carp show edema, exopthalmos and skin lesions. The requirement for fish is 3 mg/kg to 10 mg/kg of diet, and most commercial feedstuffs contain this amount. However, because of variation and uncertainty of content in feedstuffs and is processed, stored feeds, vitamin B_6 supplementation in commercial fish feeds is practiced.

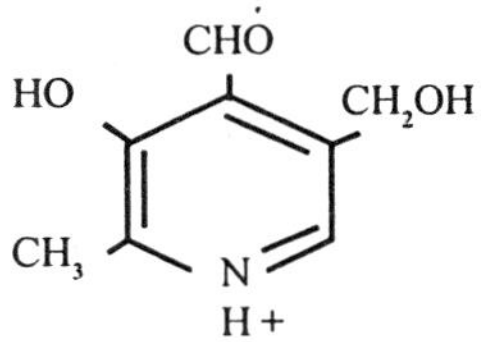

CHO, HO, CH_2OH, CH_3, N, H+

Pyridoxal

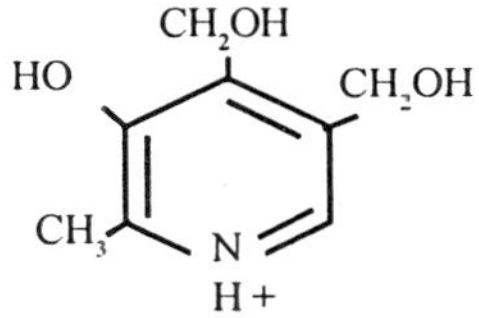

CH_2OH, HO, CH_2OH, CH_3, N, H+

Pyridoxal

CH_2NH_2, HO, CH_2OH, H_3C, N, H+

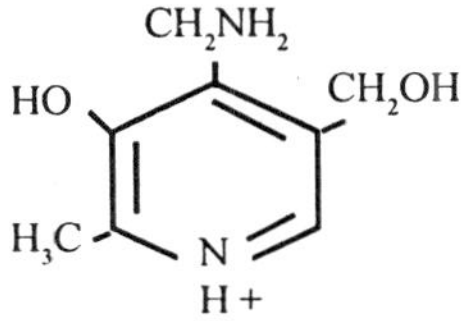

Pyridoxamine

BIOTIN

Biotin was first recognized in its association to "egg-white injury" in rats because typical deficiency signs resulted when raw egg white was supplemented into most diets. When liver was fed, egg white injury was prevented. The preventive and causative factors were subsequently identified as biotin and avidin, the latter a heat labile protein in egg white that combines with biotin and renders it inactive. Biotin functions as a coenzyme in carboxylation reactions. Enzyme systems containing biotin are acetyl-CoA carboxylase, propionyl-CoA carboxylase and pyruvate carboxylase metabolic functions requiring biotin are many and include synthesis of fatty acids, oxidation of energy yielding compounds, synthesis of purines and deamination of certain amino acids. A reliable indicator of biotin status that has been used in fish and warm blooded animals is pyruvate carboxylase or acetyl-CoA carboxylase activity.

Bioavailability of biotin in many feed stuffs is limited. For example one-half or less of the total biotin in wheat, barley, sorghum, meat and bone meal, and fish meal is available to chickens. Biotin availability in corn and soyabean meal, however, is higher, channel catfish do not require supplemental biotin in practical corn-soybean meal or corn-soybean meal-

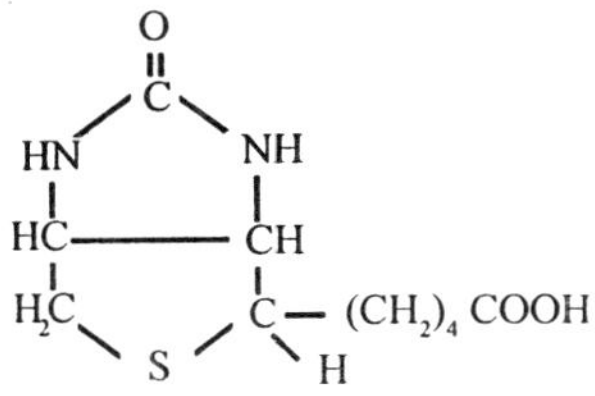

O, C, HN, NH, HC, CH, H_2C, S, C — $(CH_2)_4$ COOH, H

Biotin

fish meal diets for normal growth and pyruvate carboxylase activity. Rainbow trout also do not benefit from biotin supplementation of practical diets. The dietary requirement for fish is low, about 0.25 mg/kg of diet, so practical fish diets do not usually need supplemental biotin.

FOLIC ACID

When chemically identified, folic acid was named pteroylglutamic acid. As seen in the figure, it is composed of glutamic acid (right) plus pteroic acid, the latter being of paraminobenzoic acid and a pteridine nucleus. A number of biologically aciotive forms or derivatives of folic acid exist in nature. These include tetrahydrofolic acid, which is the active coenzyme form; 5-methyl-tetrahydrafolic acid; folic acid glutamates; and others. The folic acid coenzyme is functional in the transfer of single-carbon units, a role analogous to that of pantothenic acid in the transfer of two—carbon units. The one - carbon units may be formyl, methyl hydroxymethyl or others. Among the reactions involving tetrahydrofolic acid are the synthesis of purines and pyrimidines for formation of nucleic acids.

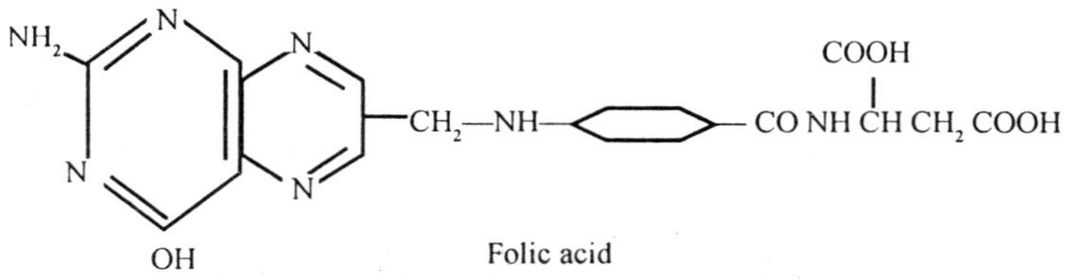

Folic acid

In animal species where folic acid deficiencies occur, megablastic anemia is usually found, characterized by large, immature erythrocytes. Megablastic anemia has been produced in salmonids, characterized by reduced production of erythrocytes, large and segmented erythrocytes with constricted nuclei, and abnormally large proerythrocytes in the erythropoietic tissues.

Folic acid deficiency could not be demonstrated in common carp. Poor growth was seen in folic acid deficient eel and channel catfish, combined deficiencies of folic acid and vitamin B_{12} enhance anemia in fish. Folic acid is synthesized by plants and microorganimsms. Grains, oilseed meals, and animal by products are all good sources. The needs of farm animals are readily met with practical needs. One to five milligrams per kilogram of diet is sufficient for young salmonids.

VITAMIN B_{12} (CYANOCOBALAMIN)

This was the last of the 15 recognized vitamins to be identified, it is also the most chemically complex. For many years it was known that liver contained a factor that would correct pernicious anemia. Supplementation of all plant diets with animal by-products provided a missing growth factor for chickens. In 1948 the factor was isolated from liver and chemical identification was completed in 1955. The molecule, which has a molecular weight of 1,354, has a cobalt nucleus in a tetra-ring porphyrin structure. Several similar compounds, where the cyanide ox nucleotide is substituted, that have B_{12} activity are found in nature; collectively, they are called cobalamins.

Vitamin B_{12} functions as a coenzyme in a number of metabolic reactions. A specific function is to act in concert with folic acid in the transfer of single-carbon units, such as methylaction of uracil to form thymine in DNA synthesis and in methyl transfer in methionine synthesis. Vitamin B_{12} is necessary for normal growth, maturation of erythrocytes, and healthy nervous tissue. A B_{12} deficiency could cause a folic acid deficiency because it is necessary for conversion of tetrahydrofolic acid to its coenzyme form. Intrinsic factor, a mucoprotein in the digestive tract, is necessary for proper absorption of B_{12}. It has been found in the gut of most animals. Vitamin B_{12} deficiency causes pernicious anemia in humans, characterized by macrocytic anemia and nervous disorders. In other animals there is usually microcytic or normocytic anemia and suppressed growth. In fish, salmonids showed microcytic anemia and fragmented erythrocytes with low haemoglobin values. Anemia was not found in B_{12} deficient channel catfish or carp, but slight reduction in growth occurred in the catfish. Evidence has been presented that intertinal synthesis is a significant source of B_{12} for channel catfish. Vitamin B_{12} is assumed to be synthesized only by microorganisms. It is a metabolic essential for all animals. Production of a dietary deficiency has been difficult in some monogastric animals; intertinal synthesis was assumed to be the camb. The enzyme glutathione peroxidase, which catalyzes the removal of metabolic peroxides.

CHOLINE

Choline has no known coenzyme fuction, but does have several metabolic roles. Its three methyl groups ($-CH_3$) in the molecule make it an important methyl donor. It reacts with acetyl coenzyme A to form acetylcholine, the neurotransmitter. It is a component of lecithin. Choline

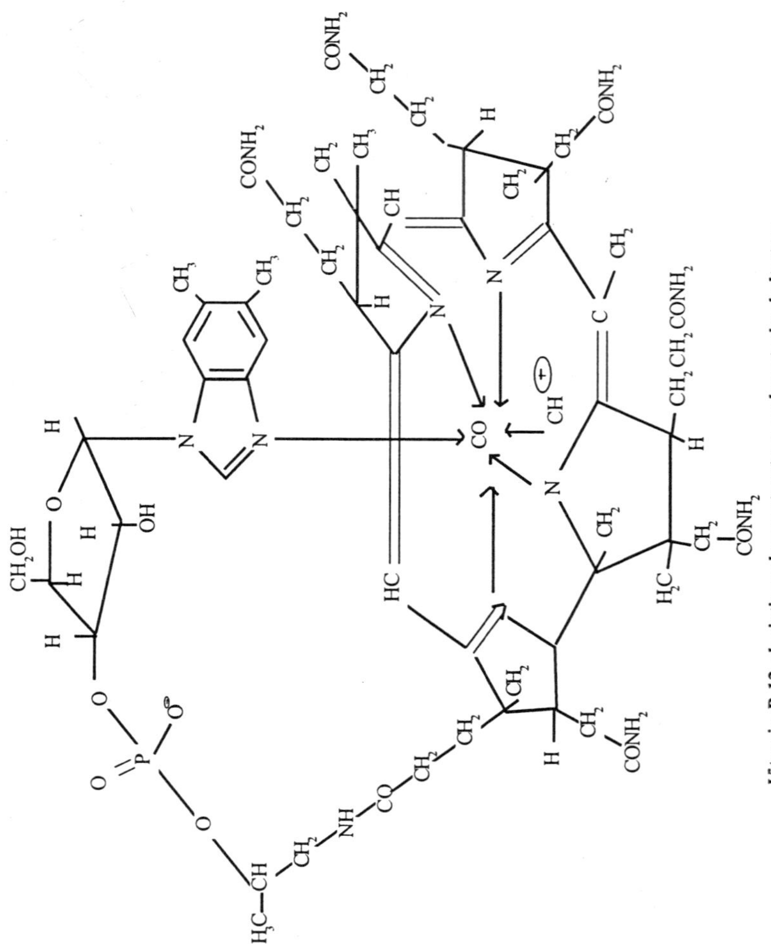

Vitamin B 12, depicting planar structure and central cobalt atom

can be synthesized in the body if "labile" methyl groups are available. Methionine or cystine may contribute methyl groups to ethanolamine to form choline. Although synthesis in the body is usually not fast enough to meet the choline requirement for normal growth, the dietary content of methionine or cystine as methyl donors or of folic acid and vitamin B_{12} for *de novo* synthesis of methyl groups can influence the dietary requirement of choline. Choline deficiency has been produced in most animals with the exception of humans. A common deficiency sign is fatty livers. All fish sps evaluated require choline in the diet. In addition to fatty livers, channel catfish had haemorrhagic kidneys and intestines, and eels had white-gray colored intestines. Choline is widely found in plant seeds. This fact, plus the ability of animals to synthesize it should limit the need to supplement fish feeds with choline. However, feeding fish feeds that contain fat extracted oilseed as a major ingredient may require choldine supplementation because choline is removed with the fat. High methionine diets will reduce the need for dietary choline.

$$(CH_3)_3N(CH_3)CH_2CH_2OH$$

Choline

VITAMIN C (Ascorbic Acid)

Finfish as well as shell fish have been found to be highly sensitive to dietary deficiency of vitamin C, especially the young fish. Gross signs, such as deformed spinal column, distorted gill support cartilage, haemorrhagic areas under the skin, depigmentation, and slow wound healing, have been produced in several fish species by feeding vitamin C deficient diets. Most animals can synthesize vitamin C in sufficient quantity for normal growth and function, but a few, such as primates, guinea pig, some birds, and many fishes cannot because they lack the enzyme L-gluco oxidase for synthesis of vitamin C from glucose. The vitamin occurs in two forms, a reduced form (ascorbic acid) and oxidized form (dehydoascorbic acid). The reduced form predominates, but the forms are biologically reversible, so both have vitamin C activity. If the dehydro-form is further oxidized to diketo-glutamic acid, it loses its activity and the reaction is irreversible. Vitamin C is a strong metabolic reducing agent. Its role in hydroxylation of proline and lysine to the hydroxy-amino acids for the conversion of procollagen to collagen has long been recognized.

Many of the deficiency signs in fish are related to malsynthesis of collagen, which is a component of bone, gill support cartilage, blood vessels, skin, fins, and wound scar tissue. Vitamin C has many other metabolic roles. Bone calcification is impaired when vitamin C is deficient. It is necessary in iron metabolism, probably to convert transferring iron from oxidized to reduced form for metabolic transport. It is required in conversion of folic to folinic acid. It is required in tyrosine metabolism; a deficiency causes tyrosine excretion in the urine. Vitamin C deficiency increases blood clotting time. It can spare vitamin E in reducing peroxidation of lipids cellular and subcellular membranes. It is essential for maximum rate of immune responses, and has a role in detoxification of various xenobiotics.

Curvature of the spinal column is a prominent, early sign of vitamin C deficiency in fin fishes. Scoliosis and lordosis (lateral and vertical curvature of spinal column, respectively) have been produced by feeding vitamin C deficient diets to rainbow trout, brook, trout, coho salmon, Tilapia, channel catfish, and young carp. Lim and Lovell (1978) described the pathology of vitamin C deficiency syndrome in channel cat fish as deformed spinal columns external and internal haemorrhages, erosion of fine, depigmented vertical bands around the midsection, distorted gill filament cartilage, and reduced rate of wound healing. Deformed head and gill operculums occur in rainbow trout deprived of dietary vitamin C. Lightner *et al.* (1979) demonstrated that without sufficient vitamin C in the diet, penaeid shrimp diet of "black death", a condition characterized by melanized haemocytic lesions distributed throughout the collagenous tissue.

Dietary requirement of vitamin C varies with metabolic function. Hilton, Cho and Slinger (1978) found that 20 mg of vitamin C per kg of diet was sufficient for normal growth in rainbow trout, but 40 mg/kg was necessary to prevent gross deficiency signs. Li and Lovell (1984) found similar results with channel catfish. Halver, Ashley and Smith (1969) reported that 50 mg/kg (the lowest level fed) was sufficient for normal growth and bone development in coho salmon, but 400 mg/kg were required for maximum rate of wound healing. Sub lethal levels of various pesticides in water increase the requirement of vitamin C by several fish. Increasing dietary vitamin C reduced incidence of vertebral damage and concentration of toxaphene in the fish, indicating that vitamin C is a factor in detoxification of toxaphene.

Dietary requirement for vitamin C by fish seems to decrease with age. Sato, Yoshinaka and Yamamoto (1978) found that young trout (6 weeks) fed vitamin C-free diets grew slowly and developed scurvy.

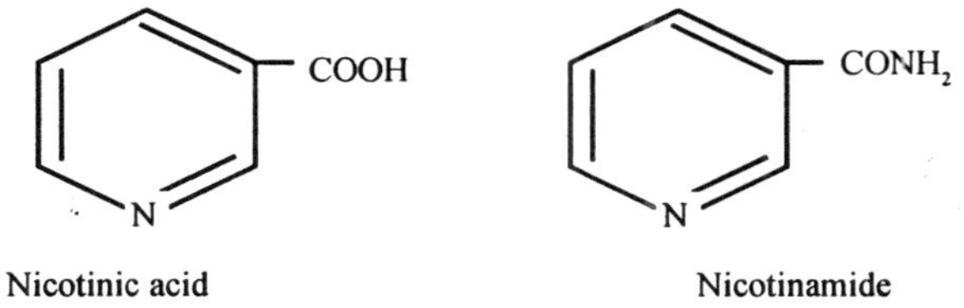

Nicotinic acid: pyridine ring with COOH

Nicotinamide: pyridine ring with $CONH_2$

Fig.: Niacin

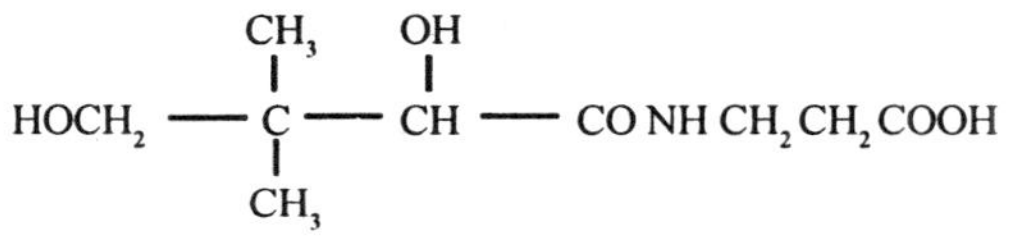

$$HOCH_2 - C(CH_3)_2 - CH(OH) - CO\,NH\,CH_2CH_2COOH$$

Fig.: Pantothenic acid

Pyridoxal: CHO, HO, CH_2OH, CH_3, N+ H

Pyridoxol: CH_2OH, HO, CH_2OH, CH_3, N+ H

Pyridoxamine: CH_2NH_2, HO, CH_2OH, CH_3, N+ H

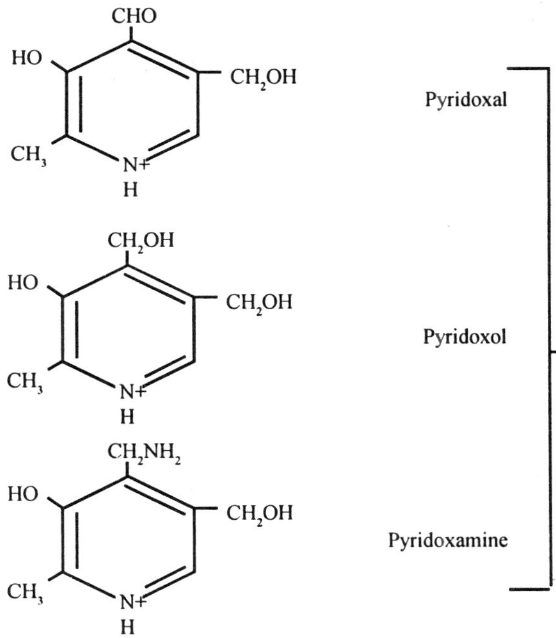

Fig.: Pyridoxine

Whereas older trout (19 months) developed none of these problems. Li and Lovell (1984) found that 60 mg/kg of diet was required for normal growth and bone requirement for small (10 g) channel catfish, but 30 mg/kg was sufficient for larger cat fish (50 g). Rainbow trout can use ascorbate-2-sulfate as a dietary replacement for ascorbic acid; channel catfish can use ascorbate - 2-sulfate, but not as efficiently as trout. Tucker and Halver (1984) contend that ascorbate-2-sulfate is a storage metabolite of ascorbic acid in fish which functions to regulate the size of tissue ascorbic acid pools interconversion of ascorbic acid to ascorbate-2-sulfate is catalyzed by the enzyme ascorbic acid-sulfatase, which is modulated by tissue levels of ascorbic acid through feedback inhibition. Guerin (1986) found, however, only small amounts of ascorbate-2-sulfate in channel catfish, and these concentrations were independent of diet or tissue levels of vitamin C. Fish fed diets deficient in vitamin C have reduced resistance bacterial diseases. Elevated dietary doses of ascorbic acid increases resistance of channel catfish to bacterial infections. When channel catfish were fed diets containing 0 mg to 3,000 mg of ascorbic acid per kilogram of diet for 13 weeks and subsequent infected with the pathogenic bacterium, Edwardsiella ictaluri, mortality over 8 days ranged from 100%. With the ascorbic acid-deficient diet to none at the highest level of ascorbic acid. Thirty mg per kg of vitamin was sufficient to prevent scurvy in the fish and 150 mg/kg significantly reduced mortality. The immune responses, antibody production, and complement activity were significantly suppressed by omission of ascorbic acid from the diet and significantly enhanced when ascorbic acid was increased from 300 mg/Kg to 3,000 mg/kg. Tissue levels of vitamin C plateau at dietary vitamin cocentrations of 500 mg/kg to 1,000 mg/kg, so maximum resistance to bacterial infection is probably provided within this diet concentration range. Raibnbow trout also showed increased antibody production when dietary vitamin C was increased to 1,000 mg/kg, which is 10 times the normal requirement. The reason that pharmacologic doses of vitamin C enhance immune responses in fish is not presently known. Commercial feed ingredients are almost completely devoid of vitamin C so, the vitamin must be supplemented into practical feeds. Vitamin C as 1-ascorbic acid is highly sensitive to oxidative destruction during processing and subsequent storage of feeds. Approximately 25% of 1- ascorbic acid is destroyed during steam pelleting and 50% is lost in extrusion processing; thus, over fortification is necessary unless the vitamin is added after processing. The half-life of 1-ascorbic acid in fish feed during past-processing storage under warm weather conditions is approximately 2.5 months. Phosphate and sulfate conjugates of ascorbic acid are much more stable against oxidation during processing and storage than 1-ascorbic acid. These compounds have vitamin C activity for fish and will likely replace the less stable form.

INOSITOL

It is widely distributed in plant and animal tissues. In animals it occurs freely as myoinositol, ox as a component of phospholipids as

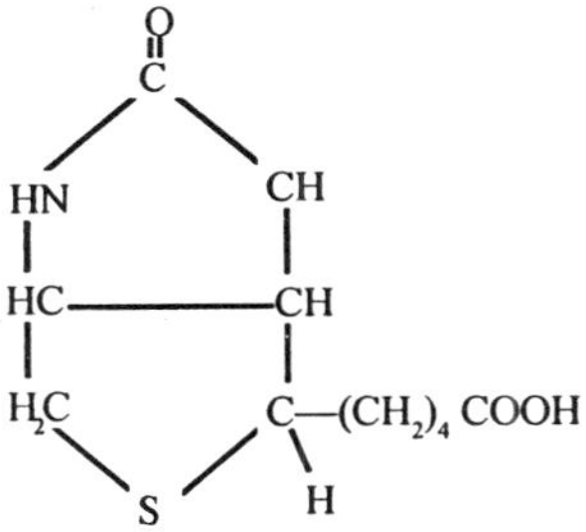

Biotin

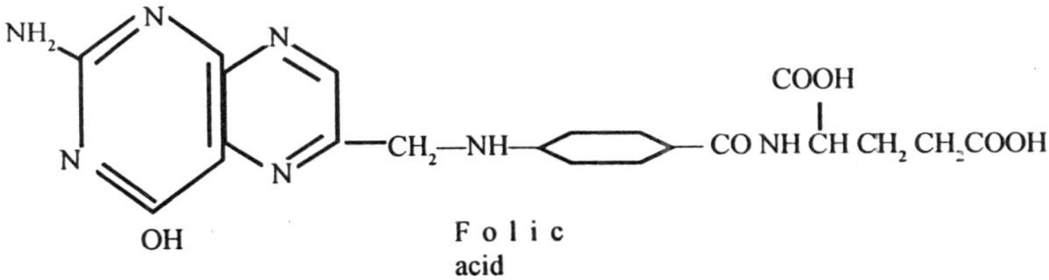

Folic acid

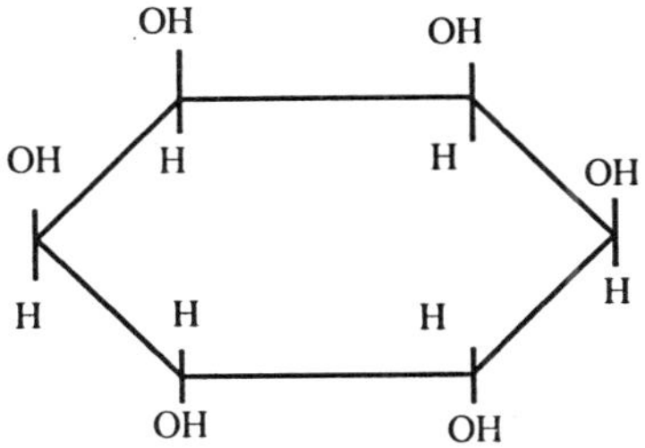

Inositol

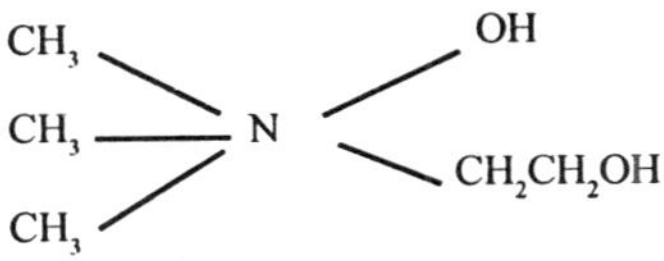

Choline

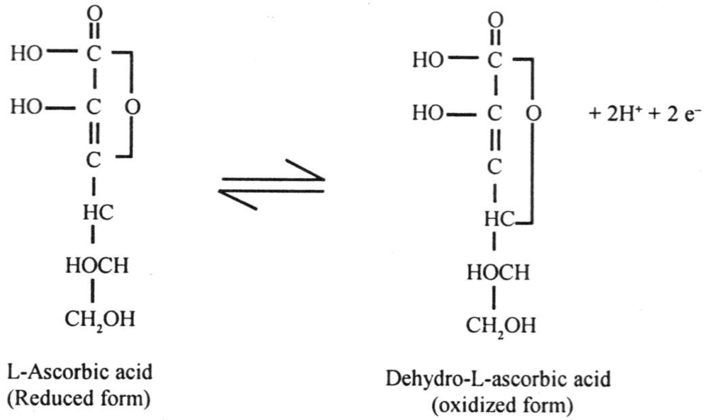

O
||
HO— C
|
HO— C O
||
C
|
HC
|
HOCH
|
CH_2OH

L-Ascorbic acid
(Reduced form)

⇌

O
||
HO— C
|
HO— C O
||
C
|
HC
|
HOCH
|
CH_2OH

$+ 2H^+ + 2e^-$

Dehydro-L-ascorbic acid
(oxidized form)

Ascorbic acid

lipositol, in cell membranes. In plants, it is most concentrated in seeds and about two-thirds of it is in the form of a hexaphosphate ester, phytic acid, which is highly indigestible to fish and other monogastric animals. Inositol has no known coenzyme function. Besides being a component of cell membranes, it apparently has some lipotropic action. Rainbow trout fed inositol-deficient diets had large accumulations of triglycerids and cholesterol but low levels of phospholoipids in the liver. In several fish species (trout, prawn, sea bream, carp and eel), reduced growth, anemia, fin erosion, and slow rate of gut emptying have been reported with inositol deficiency. Inositol deficiency is difficult to produce in some animals when the diet is nutritionally complete otherwise.

FAT SOLUBLE VITAMINS

VITAMIN A

Vitamin A is found only in the animal kingdom. It exists in the free alcohol form as retinol and as esters of higher fatty acids. One international unit (IU) of vitamin A is equal to 0.3 ug of all transretinol. Vitamin A_1 $C_{20}H_{30}O$ has been isolated from lipids of many fish and land animals, where as vitamin A_2 has been isolated only from freshwater fish. Plants produce red-to-yellow pigmented compounds called carotenoids some of which have vitamin A activity. Beta-carotene has by far the highest vitamin A activity. This compound appears to be capable of yielding two moles retinol upon simple hydrolytic cleavage; however, the vitamin A activity of beta-carotene is much less than this for most animals. Several fishes have been found capable of using beta-carotene for vitamin A activity; however, Poston *et al.* (1977) found that cold water fish could utilize beta-carotene at 14c Lee (1987) found that channel cat fish readily converted beta-carotene into vitamin A and A_2 in almost 1 to 1 ratio.

Vitamin A, like other fat soluble vitamins, is stored in large amounts in the body (liver) if intake exceeds metabolic need. It is possible for fish to store enough vitamin A in its role to produce a toxic condition (hypervitaminosis); however, prolonged consumption of a diet with an unusually high level of vitamin A would be required to produce this. An established physiological function of vitamin A is its role in vision. Retinol is combined with a protein, opsin, to form rhodopsin, which is the compound involved in the photochemical reaction in the retina of the eye in the process of vision. Other metabolic roles of vitamin A are less well developed understood. It is used for maintenance of mucosal membranes that line many body organs, the gastrointestinal tract, the respiratory tract, and the eye.

In vitamin A deficiency, epithetial cells fail to differentiate beyond the squamous type to mucus-secreting type, and mesenchyme cells fail to differentiate beyond the blast stage. Epithelial cells from the eye and many other areas of the body keratinize and this lowers resistance to infection. Several deficiency symptoms appear to be related to impaired function of epithelial tissue. Reproduction is impaired in most animals. Vitamin A deficiency in salmonids has been described as reduced growth rate, light skin colour, ascites (fluid in abdominal cavity) and pathological condition of the eye characterized by exopththalmos, haemorrhagic eyes, eye lense displacement, thinning of cornea, degradation of the retina, and twisted gill opercula. Channel catfish fed vitamin A deficient diets over a long period (2 years) developed exophthalmos, edema (collection of fluid in tissues) and kidney haemorrhage. Common carp showed dificiency signs of light skin colour, fin and skin hemorrhages, exophthalmos, and deformed gill opercula. Retinol and beta-carotene are sensitive to oxidation, so natural sources of vitamin A may be oxidized to various degrees. Thus supplemental vitamin A should be added to fish feeds. Synthetic vitamin A used in fish feeds is in stabilized forms, usually as palmitate, acetate or propionate esters. Dry additives are usually in beadlet form. Where the retional ester is coated with gelatin or some other oxygen barrier.

VITAMIN D

Several biologically active forms of vitamin D occur and the chemical structure of one, vitamin D_2 or ergocalciferol, with the chemical formula $C_{28}H_{44}O$. Vitamin D_3 or activated 7-dehydrocholesterol has the

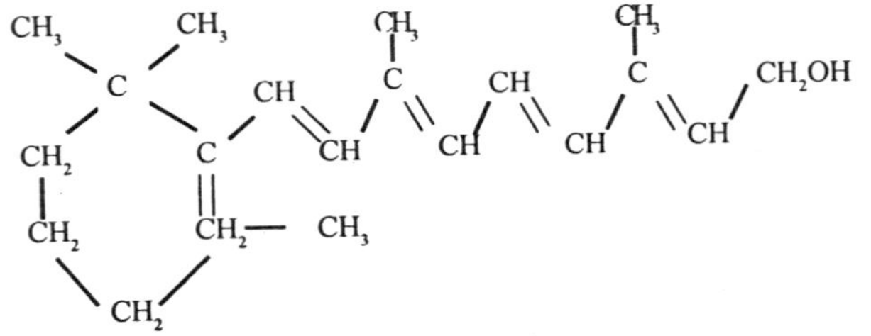

Retinol

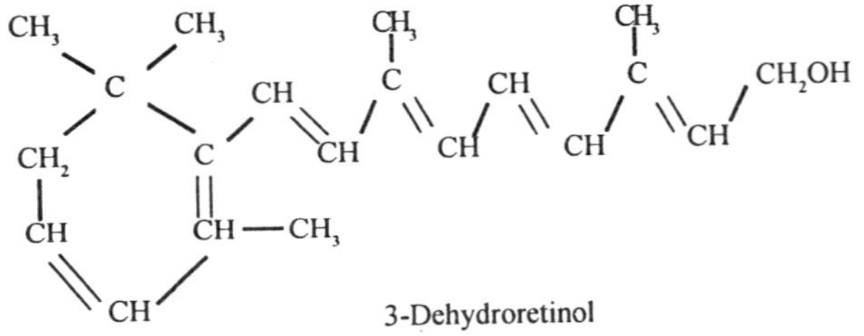

3-Dehydroretinol

chemical formula $C_{27}H_{44}O$ and contains more simplified, unsaturated eight carbon side chain. Vitamin D_3 (Cholecalciferol) is formed in most animal tissue by rupture of one of the ring bonds of 7-hydrocholesterol when exposed in the skin to ultraviolet radiation.

Vitamin D_3 functions as a precursor of 1,25-dihydrocholecalciferol, which stimulates the absorption of calcium from the intestine. Vitamin D is essential for maintaining homeostasis of calcium and inorganic phosphate. Vitamin D is involved in alkaline phosphatase activity, promotes intestinal absorption of calcium and influences the action of parathyroid hormone on bone.

VITAMIN E

Vitamin E is present in at least eight tocopherols which occur in plants. The name tocopherol is from the Greek word tocos, which means childbirth. Alpha toco has the highest vitamin E activity. Vitamin E activity of compounds is measured in International units (IU) with 1 IU being equivalent to the biological activity of 1 mg of D- α-tocopherol (replacing the formerly used D1α-tocopherol acetate). Traditionally biological activity has been based on prevention of fetal resorption in rats. A major function of vitamin E is its role as a metabolic antioxidant, with a specific role in preventing oxidation of unsaturated phospholipids in cellular membranes, such as erythrocytes, and subcellular membranes such as

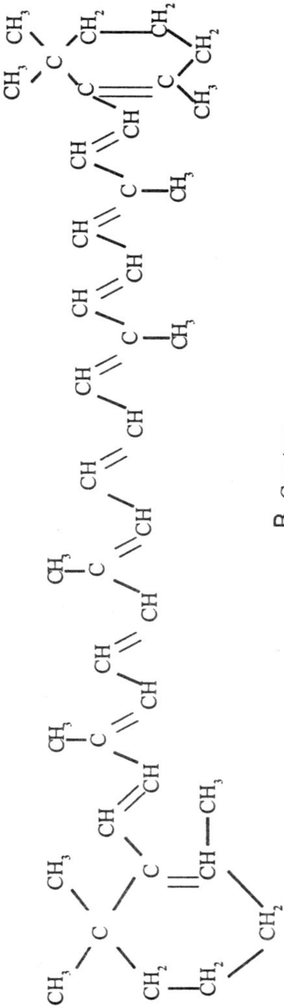

β-Carotene

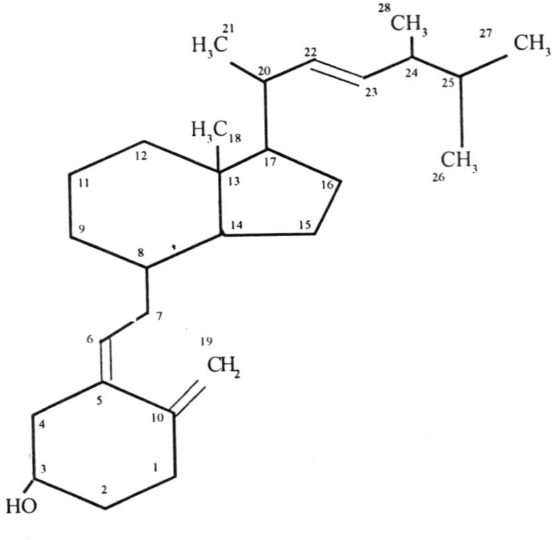

Vitamin D_2

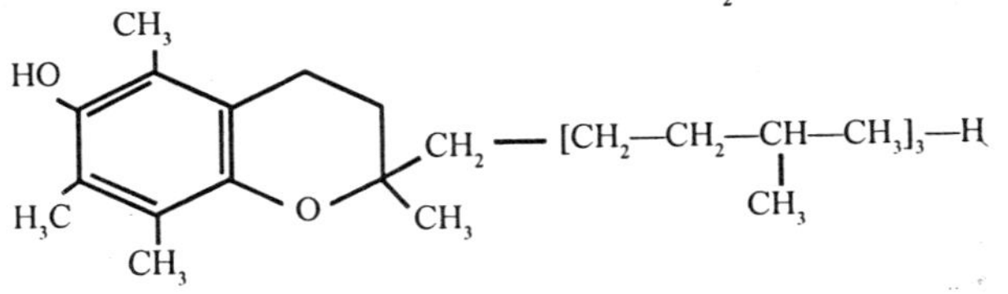

α-Tocopherol (5,7,8-Trimethyltocol)

mitochondria. It is often referred to as a metabolic free radical scavenger or peroxide scavenger. In most animal species, an increase in polyunsaturated fatty acids, especially when partially oxidised, in the diet produces an increase in dietary need for vitamin E.

A specific role for vitamin E is an enzyme system has not been identified, although impairment of several enzyme systems, such as those involved in porphyrin and heme synthesis, has been identified in vitamin E deficiencies in various animals. Diverse physiological abnormalities have been demonstrated with vitamin E deprivation in animals. Common are nutritional muscular dystrophy, which has been produced in several fish and terrestrial animals and pathological conditions of male and female reproductive organs, causing reduced fertility and reproduction. Increased permeability of capillaries, which result in haemorrhages and edema in various body areas, has been caused by vitamin E deficiency in various animals. The syndrome exudative diathesis, manifested by accumulations

Vitamin E

of fluid under the skin or in the abdominal cavity (sometimes of a greenish colour caused by decomposed haemoglobin), has been produced in channel catfish and salmonids fed vitamin E deficient diets. Vitamin E deficiency causes reduction in ability of erythrocytes to withstand peroxide deterioration of membranes. Severe anemia, characterized by immature, irregularly shaped and sized erythrocytes, is produced in vitamin E deficiency.

Other vitamin E deficiency signs described for several fish species include fatty livers and ceroid (dark lipoid) bodies in liver. "Sekoke disease" in common carp characterized by thinning of flesh on the back of fish, was caused by feeding oxidized silk worm pupae, but corrected by supplementing the diet with vitamin E. Subclinical measurements used to detect vitamin E deficiency include erythrocyte fragility and histological examination of tissues for necrosis of muscle fibres and ceroid concentration in liver and kidney.

In most experimental and practical situations where the classical vitamin E deficiency syndrome in fish has been produced, inclusion of high levels of polyunsaturated fatty acids or omission of selenium from the diet has been necessary. However, Lovell, Miyazaki and Rebegnator (1984) fed channel catfish diets low in polyunsaturated fatty acids, using stripped lard as the lipid source, and produced fish with severe nutritional muscular dystrophy and other signs of vitamin E deficiency. This implies that fish feeds not containing high levels of polyunsaturated lipids, such as many commercial warm water fish feeds, can cause reduced growth rate and various pathologies when deficient in vitamin E.

VITAMIN K

The name vitamin K was given to a fat-soluble factor necessary in the diet of chicks to prevent haermorrhage and for normal blood clotting—several compounds with Vitamin K activity has been isolated or synthesized. These include phylloquinone (vitamin K_1) found in green plant leaves; and menaqunone (vitamin K_2) isolated from fish meal and animal feces.

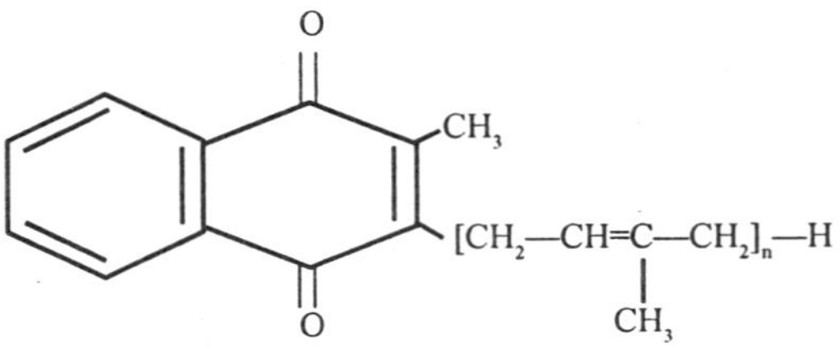

O

CH_3

$[CH_2—CH=C—CH_2]_n—H$

CH_3

O

Vitamin K_2
(n may be equal 6,7 or 9 isoprene units)

Vitamin K is necessary for normal blood clotting in all animals, including fish. Several proteins necessary for blood coagulation are dependent upon vitamin K for their synthesis. These include prothrombin, proconvertin, plasma thromboplastin antecedent, and stewart-power factor. The role of vitamin K is assumed to be in conversion of the precursor to the factor (prothrombin to thrombin) by the carboxylation of the glutamic acid residues to form r-carboxyglutamic acid in the active proteins. Other proteins, such as in bone and kidney have been found that contain r-carboxyglutamyl acid residues, so it is presumed that vitamin K is involved in other carboxylase enzyme systems besides blood clotting. Channel catfish and trout require dietary vitamin K for normal blood coagulation. Growth rate was not afffected in either of the fish when vitamin K was deleted from the diets. Quantitative requirement for vitamin K has not been determined for warm water fishes, levels of 5 mg to 1mg of menadione per kg of diet is sufficient to maintain normal blood coagulation time in trout. Fish meal and alfalfa meal of a good sources of vitamin K. Menadione sodium bisulfite or menadione dimethyl pyrimidinal bisulfite are synthetic sources used in commercial feeds; the latter is more heat stable during feed processing.

SOURCES OF VITAMINS

Vitamins most likely to be deficient in commercial fish feeds that contain oilseed meals, animal by products and grains are vitamins C, A, D, Niacin, pantothenic acid, riboflavin, and possibly vitamins E and K. inositol, biotin, folic acid, pyridoxine and thiamin are widely found in plant feedstuffs and vitamin B_{12} is present in animal byproducts. Vitamin E and choline are usually extracted with the oil or germ from plant seeds, and may be deficient in many commercial feed formulations. It is possible to formulate a fish feed from commercial ingredients that is adequate in all essential vitamins except vitamin C. Brewer's yeast or distillary dried grains are good sources, and grain by products that include the seed coat are fair sources of the water. Soluble vitamins (except C); alfa alfaya meal is a good source of vitamin E and K; fish meal is a good source of B_{12}; and fish liver oil can supply A and D.

INTESTINAL SYNTHESIS OF VITAMINS

Microbial synthesis of vitamins in the gut has not been researched well in fish, although significant amounts of B_{12}, folic acid, inositol, biotin, and K are attributed to this source in warm blooded animals studies conducted at Auburn university in which the ratios of vitamins to indigestible dry matter (IDM) were measured in the diet and in the feces (removed from rectum) showed significant increases in inositol and vitamin B_{12} in the digestive tract of channel catfish and tilapia. There was no increase in biotin. Addition of antibiotics significantly reduced the ratio of vitamin to IDM in feces. Increase in vitamin B_{12} in the digestive tract of Nile tilapia was about 5 times higher than in the digestive tract of channel catfish. Significant absorption of intestinally synthesized B_{12} by channel catfish was revealed by feeding the fish ^{60}Co in the diet and recovering radio-labeled vitamin B_{12} in the blood, liver, kidneys and spleen.

MINERALS

Not all inorganic elements found in any animal body are essential in its diet. However, dietary need for 22 minerals has been demonstrated in one or more animal species. Those required in large quantities are termed "major" and those required in trace quantities are called "trace" minerals. The major minerals are calcium, phosphorus, magnesium, sodium, potassium, chlorine and sulfur. Trace minerals are iron, iodine, manganese, copper, cobalt, zinc, selenium, molybdenum, fluorine, aluminium, nickel, vanadium, silicon, tin and chromium. The mineral requirements of fish have been studied only sparsely. In addition to the general problems encountered in mineral nutrition research, such as formulating mineral free diets and overcoming tissue stores of minerals, fish absorb dissolved minerals from the water. A major difference between mineral metabolism in fish and land animals is osmoregulation or maintenance of osmotic balance between body fluids in the fish and the water around the fish. Other biochemical functions of minerals in fish are similar to those in warm blooded animals. Some minerals are constituents of hard tissues such as bone, fins and scales and some are components of soft tissues, such as sulfur in protein and iron in hemoglobin. Some minerals function as components or activators of enzymes and hormones, such as zinc, which activates alkaline phosphatose and iodine, which is a component of the hormone thyroxine. Some soluble elements, such as calcium, sodium, potassium and chloride have functions in the blood or body fluids such as osmoregulations, acid-base balance, and sensitizing muscle fibres.

TABLE-9.5 : VITAMIN DEFICIENCY SYMPTOMS IN FISH

Vitamin	*Deficiency symptoms*
Thiamin	Anaemic, convolutions, corneal opacities, fatty liver, loss of equilibrium, melanosis in older fish, muscle atrophy, paralysis of dorsal and pectoral fins, rolling, whirling motion, weakness.
Riboflavin	Anorxia, cloudy lens, darkened skin, dim vision, discoloured iris, haemorrhage in eyes, incoordination, lens cataract, mortalities, photophobia, xeropthalmia.
Pyridoxine	Anorxia, ataxia, convolutions, flexing of opercles, hyperexcitability, rapid jerky breathing, spasms, weight loss, nervous disorders, increased mortalities.
Folic acid	Anaemia, dark colouration, exophthalmia, lethargy, pale gills.
Pantothenic acid	Anorxia, dark colouration, lethargy, necrosis of jaw, poor weight gain.
Inositol	Anaemia, bloated stomach, poor growth.
Biotin	Blue slime disease, mortalities, muscle atrophy, dark colouration.
Choline	Anaemia, poor food conversion, poor growth.
Nicotinic acid	Anaemia, edema of stomach and colon.
p-Benzoic acid	No significant change in growth.
Cobalamine	Erratic haemoglobin and erythrocyte counts.
Ascorbic acid	Impaired collagen production, impaired wound healing.
Vitamin A	Edema, exopthalmos.
Vitamin D	Reduced conversion.
Vitamin E	Anaemia, mortalities.

REQUIREMENTS

Fish can absorb dissolved minerals from the water across the gill membrane or, in the case of marine fishes that drink water, through the digestive tract. Most of the calcium requirement for fishes comes from the water. In sea water, significant amounts of iron, magnesium, cobalt, potassium, sodium and zinc can be obtained from the water. Berg (1968) provided data that indicated that gold fish obtained 50% to 80% of their calcium from the water when fed a calcium-adequate diet. Fish required a dietary source of phosphorus to meet their relatively high metabolic requirement because levels dissolved phosphorus in natural waters are relatively low.

CALCIUM AND PHOSPHORUS

Most of calcium found in the fish body, perhaps 99% is in skeletal tissue and scales. From 20% to 40% of the total calcium is in scales. During fasting calcium is reabsorbed from the hard tissues for physiological functions. The percentage of calcium in the whole, fresh (wet) body of fin fish ranges from 0.5% to 1% with a ratio of calcium to phosphorus of 0.7 to 1.6. In addition to its structural functions in bones and scales, calcium, is essential for blood clotting, a muscle function, nerve impulse transmission, osmoregulation and as a cofactor during various enzymatic processes. Approximately 85% to 90% of the phosphorus in fish is in bone and scales. In bone, phosphorus is complexed with calcium to form apetite, or tricalcium phosphate. Phosphorus also functions in a variety of organic phosphates, such as constituents of adenosine triphosphate (ATP), deoxyribonucleic acid (DNA), ribonucleicacid (RNA), various coenzymes and phospholipids in cell and subcellular membranes. The phosphate buffer system maintains normal pH in body fluids. Calcium deficiency was produced in channel catfish grown in calcium free water. This was characterized by reductions in growth and ash content of bones. Signs of dietary phosphorus deficiency for most fishes include poor growth and bone mineralilzation. Further signs of deficiency observed in carp include increases incareass fat; reduced blood phosphate levels, deformed head (frontal bone), abnormal calcification of ribs and soft rays of the pectoral fins, and a curved or abnormal spine. Symptons of phosphorus deficiency in red sea bream include deformed vetebrate and increased fat content and decreased glycogen content in the liver. Minimum requirement of available phosphorus in diets for channel catfish was determined to be 0.45% with purified diets. Requirements for common carp, Nile tilapia, red sea bream, and eel have been determined to be 0.6%, 0.9%, 0.68% and 0.29% of diet, respectively. Phytate phosphorus (approximately 67% of the phosphorus in grains is in the phytate form) is poorly available to fish. Phosphorus in fish meal is 40% to 75% available to fish with gastric stomachs, but less than 25% available to the stomachless carps. Inorganic phosphorus from sodium or monocalcium phosphate is highly available to all fish, however, dicalcium phosphate is less available.

MAGNESIUM

About 70% of the magnesium in a fish's body is in the hard tissue other functions of magnesium are as an enzyme activator in carbohydrate metabolism and in protein synthesis. Magnesium is necessary in body fluids to maintain integrity of smooth muscle. Deficiency causes tetany in warm blooded animals and flaccid muscle in fish.

Magnesium deficiency in the diet of channel catfish causes poor growth, anorxia, lethargy, flaccid muscles, high mortality and depressed magnesium levels in the whole body, blood serum and bones. Deficiency of magnesium in the diet of common carp caused similar signs, in addition to convulsions, signs of deficiency in rainbow trout include mortality, vertebral curvature, degeneration of muscle fibres, and degeneration of the epithelial cells of the pyloric caecae and gill filaments. Red sea bream rared in sea water, which typically contains high levels of magnesium, showed no signs of deficiency. Fish in freshwater, which only contains 1 mg/L to 3 mg/L of magnesium, required 0.025% to 0.07% magnesium in the diet. Most foods especially plants, are high in magnesium, which usually makes it unnecessary to supplement feeds made from natural ingredients; however, purified diets must have a magnesium supplement.

IRON

The principal role of iron in a fish body is as a component of hemoglobin. Another role is as a component of the cytochrome enzyme system that regenerates ATP in cellular oxidation. Hemoglobin is the oxygen-carrying pigment in red blood cells. Red blood cells are formed primarily in the spleen and anterior kidney in fish instead of the bone marrow, as in land animals. Red blood cells are regenated periodically and most of the iron is recycled. That which is not recycled is excreted through the bile into the intestine. A deficiency of iron can cause anemia. Iron, like other elements of low solubility, such as zinc and copper, is absorbed and transported in the body in protein-bound form. In the intestinal cell mucosa, iron combines with a protein, apoferritin, to form ferritin. The amount of apoferritin in the mucosa is regulated by body need for iron. Iron is in oxidized form, when combined with the protein in the mucosa. When liberated into the blood, it is reduced to Fe by reducing agents, such as vitamin C. It is transported in the blood bound to another protein as transferring and stored in the liver and hemopoietic tissues as Fe^{++} in combination with a protein until used. Iron and other minerals of low solubility are not excreted through the urine, but are returned to the digestive tract. Dietary deficiency of iron causes microcytic anemia in fish. Dietary requirement of iron by channel cat fish is 30 mg/kg of diet. Dissolved iron in the water can serve as a source of iron for fish metabolism, although dissolved iron often precipitates out as ferric hydroxide and levels in solution are low, iron is side spread in feed stuffs; however, its availablility from plant feeds is relatively low unless fish feeds contain significant amounts of animal by products, supplemental iron should be used.

COPPER

Copper is involved with iron absorption and metabolism. When the diet is deficient in copper, iron levels in body tissues decrease. Copper functions in hematopoiesis (hemoglobin formation) and in several enzyme systems, such as cytochrome C oxidase and tyrosinase. It is essential in bone development. Perhaps through its role in collagen synthesis. Some sea animals, molluscs and crustaceans contain copper as the metal nucleus of the oxygen carrying pigment in the blood, hemocyanin or cyanodin, which has an analogous role to hemoglobin in red-blooded animals. Like iron, copper is absorbed and transported as a copper-protein complex. Common carp require approximately 3 mg of copper per kg of diet for normal growth. Channel catfish required 1.5 mg/kg to 5mg/kg for normal growth and blood cell formation, but 32 mg/kg caused growth depression and anemia. This indicates a rather narrow range of dietary copper tolerance. Approximately 100 mg to 250 mg copper/kg of diet is toxic to livestock.

IODINE

Iodine is a component of thyroxin, the thyroid hormone that regulates rate of metabolism. If the amino acid tyrosine and iodine are supplied, the body can synthesize thyroxin. Deficiency of iodine results in hyperplasia of the thyroid gland, or goiter, in fish. Reports have shown that iodine can be absorbed efficiently from the gut and is trtansported in the body bound to a protein. It is excreted in the urine. The minimum requirement of fish for dietary iodine has not been defined; however; 1 mg/kg to 5 mg/kg feed has been found to be an adequate level. Fish meal is a rich source of iodine fish feeds not containing fish meal should probably be supplemented with iodine in the presence of natural aquatic foods.

ZINC

Zinc has several functions. It serves as a cofactor in several enzyme systems, including carbonic anhydrase found in red blood cells, enzymes in protein digestion, and enzymes in carbohydrate catabolism. It plays a role in preventing teratinization of epithelial tissues. Insulin is stored as Zinc complex. Gatlin and Wilson (1983) demonstrated that Zinc-deficient channel catfish had depressed growth, appetite and serum alkaline phosphatase activity and reduced levels of Zinc and calcium in bones. They showed that channel catfish had a minimum dietary zinc requirement of 20 mg/kg to prevent deficiency signs, but the requirement for optimum growth was lower. Zinc deficiency in common carp caused slow growth; loss of appetite high mortality, and erosion of the skin and fins. Dietary levels of 5 mg/kg allowed maximum growth rate, but 15 mg/kg to 30 mg/kg were required to prevent all deficiency signs.

MANAGANESE

Manganese functions as a cofactor in several enzyme systems, including those involved in synthesis of urea from ammonia, amino acid metabolism, fatty acid metabolism and glucose oxidation. Manganese-deficient diets caused depressed growth in common carp and rainbow trout, and abnormal tail growth and shortening of the body in the latter. Suplementation of the diet to bring the level of managanese to 12 mg/kg to 13 mg/kg improved growth in both species and prevented abnormalities in rainbow trout.

SELENIUM

The most notable function of selenium is as a component of the enzyme glutathione oxidase, which acts along with vitamin E as a biological antioxidant to protect polyunsaturated phopholipids in cellular and subcellular membranes from oxidation damage. Selenium has also been identified as a cofactor in glucose metabolism. Selenium deficiency in Atlantic salmon caused increased mortality and suppressed plasma glutathione peroxidase activity. Either selenium or Vitamin E deficiency caused nutritional muscular dystrophy. Maximum plasma glutathione peroxidase activity occurred in rainbow trout fed diets containing 0.38 mg/kg of selenium.

SODIUM, POTASSIUM AND CHLORIDE

Dietary deficiencies of sodium, potassium and chloride have not been produced in fish, although these elements are necessary for osmoregulation and pH balance in the body fluids, nerve impulse transmissions and other functions. Sodium and potassium are the major extra cellular and intracellular cations, respectively and chloride is the major extracellular aninon. The chloride ion is a component of hydrochloric acid, which is secreted in the stomach. Most freshwater and all sea water probably contains sufficient amounts of these three ions to satisfy the physiological needs of fish. Fish absorb the ions through the gills in freshwater and through the gut in sea water. Channel catfish did not respond to the addition of salt (NaCl) to salt free purified diets fed in freshwater. Because most fish can excrete large dietary intakes of salt effectively, dietary levels of 8 to 12 per cent salt have had no adverse effect on sub adult or adult fish fed in fresh or sea water.

SUPPLEMENTARY FEEDING

Culture of a number of compatible non-predatory fish of different but complimentary food habits cultured together to make best use of all the natural food material present in the pond, gives high production. To accelerate fast growth in fishes and also to permit increased stocking rate, many kinds of supplementary feeds are used. The primary objective of feed formulation is to provide the species under culture with an acceptable diet that meets its nutritional requirements at different stages of its life, so as to yield optimum production at minimum cost. The type of feed required and the methods of processing will influence formulation. Extruded or floating type feeds must contain an appreciable quantity of starch for satisfactory gelatinisation and expansion. Most of such feeds contain 20-25% cereal grain such as corn, wheat or sorghum. The type of culture system also has a determining role in feed formulation. For example, if the feed is for use in semi-intensive pond culture, certain vitamin and mineral supplements can be omitt from the formula, as these nutrients are likely to be available to the animal from the natural food organisms growing in the ponds. Since protein is the most expensive portion of an animal diet, it is usually computed first in the diet formulation. The first step consists of balancing the crude protein and energy levels. The most commonly used for balancing crude protein levels are the square method and algebraic equations. For example, to balance a supplementary feed to conatin 25% protein using only two ingredients–fish meal (50% protein) and rice bran (8% protein)–a sqaure is constructed as shown below:

Fish meal 50% ________________ 17 parts of fish meal;

$$\frac{17}{42} \times 100 = 40.48\%$$

Rice bran 8% ________________ 25 parts of Rice bran

$$\frac{25}{42} \times 100 = 59.52$$

The desired protein level of the feed (25%) is inserted in its centre. The two feed stuffs, along with their protein content, are placed on each corner at the left hand side of the square and the levels of protein of each feed stuff are subtracted from the desired protein level of the feed. The differences are placed on the corners of the square diagonally opposite the feed stuff, ignoring the plus or minus. The difference between the percentages of protein in the rice bran and the protein required in the feed under formulation show the proportion of fish meal needed. The difference between the protein percentage of fish meal and of the feed being formulated show the proportion of rice bran required. These proportions can be expressed on a percentage basis, as 40.48% fish meal and 59.52% rice bran or as ratio of 17 parts : 25 parts.

Algebraical equations can be used to arrive at the same percentages as follows:

Assume

X = Fish meal in kg per 100 kg feed
Y = Rice bran in kg per 100 kg feed

X + Y = 100 kg feed — (1)

0.50 X + 0.08 Y = 25 kg protein per 100 kg feed — (2)
Multiply equation (1) by 0.08
0.08 X + 0.08 Y = 8 — (3)
Subtract equation (3) from (2)
0.42 X = 17

$$X = \frac{17}{0.42} = 40.48 \text{ kg or } \%$$

From equation (1) one can derive

Y = 100 - X
Y = 100 - 40.48 = 59.52 kg or %

In actual practice, more than two ingredients are generally used in feed formulations. For example, to balance a diet containing 30% protein using fish meal (51% protein), rice bran (8% protein) and corn meal (10% protein) in the proportions 2 parts soyabean meal:1 part fish meal and 1 part rice bran to 1 part corn meal.

Culture of compatible non-predatory fish of different but complimentary food habits cultured together to make best use of all the natural food material present in the pond, gives high production. To accelerate fish growth and also to permit increased stocking rate, many kinds of supplementary feeds are used. The selection of supplementary feed depends on number of factors such as:

1. Ready acceptability to cultured animal
2. Easy digestibility
3. High conversion value
4. Easy transportability
5. Abundant availability.

10

PATHOLOGY

CLINICAL EXAMINATION OF DISEASED FISH

Clinical examination is the first means of assesssing possible disease condition in finfish. Observations of normal animals are compared with those of abnormal, moribund and, to a lesser extent, dead specimens. This chapter is compilation of reported disease observations, which may prove helpful in indicating procedures to be followed. Clinical observations have to take into account environmental conditions, which include normal water temperature, fluctuations and changes in salinity and pH due to varying freshwater flows, rainfall or drought. Such factors in turn effect the water's oxygen content and subsequently have an impact over the flora and fauna of that particular area. When the clinical examination has been completed, the veterinarian or fisheries manager/scientist will often be able to suggest modifications of farming practices as a first step in dealing with the disease. These initial observations should be immediately followed up by appropriate laboratory tests, leading to a presumptive, and later a confirmatory, diagnosis, and thus to appropriate specific treatment. It is important to gather all this evidence because illness in fish is not always clearcut; the diagnosis often being complicated by secondary or opportunistic infections, multiple infections or physiological reactions to an environmental stressor which is no longer present.

MORTALITIES IN FISH POPULATIONS

Observations of fish populations may provide initial clues to the possible cause of death. The signs, which have been summarized in the Table 10.1.

TABLE-10.1 : DISEASE SIGNS REVEALED BY OBSERVATION OF INFECTED STOCK

Characteristic mortality pattern and environmental indications	*Possible cause*
Extended course of mortalities : More noticeable with confined fish	Internal and external parasites
Mortalities raise from numbers, peak, and remain at the high level	Nutritional deficiency
Mortalities raise from low levels, peak then fall again. Mainly seen on fish farms	Bacterial, Viral or Fungal infections
Mortalities occur early in the morning. Onset abrupt. Larger fish and more active fish species die initially. Often changes in colour and smell of water. Aquatic plants and algal bloom often die	Low oxygen, i.e., Hypoxia
Fish kill is abrupt. Mortalities occur at any hour of the day or night. All sizes and species of animals may be affected–fish, zooplankton, molluscs, crustaceans, amphibians. Small fish die first–usually showing loss of equilibrium and sometimes convulsions	Toxicants
Acute mortalities over an extended period, environment near hydro electric discharge or dam overflow or pumped-water fish farm	Gas bubble disease (Embolism)

SIGNIFICANT BEHAVIOURAL SIGNS OF DISEASE IN CONFINED FISH

The changes in fish behaviour attributed to disease have been summarized in Table 10.2.

TABLE-10.2: BEHAVIOURAL CHANGES IN FISH ATTRIBUTED DUE TO DISEASE OUTBREAK

Behavioural changes of fish	*Possible cause*
Many fish active at surface, gulping air	1. Lack of oxygen in water. 2. Protozoan gill parasites. *Chilodonella sps, Trichodina sps, Amoebae, Ichthyobodo necator.*

{cont.}.....

	3. Copepod parasites of gills: *Ergasilus sps, Achtheres sps; Salmincola sps.* 4. Gill trauma:Mechanical damage, Chemical damage 5. Nutritional deficiency : Pyridoxine
Many fish crowding water inlet or crowding air supply	1. Lack of oxygen in water. 2. Protozoan gill parasites. *Chilodonella sps, Trichodina sps, Amoebae, Ichthyobodo necator, Ichthyothirius multifilitis* 3. Helminth parasites–Monogenetic trematods: *Dactylogyrus sps.* 4. Gill trauma : Mechanical damage.
Fish lying restlessly near water surface with little movement, sometimes gulping air.	1. Protozoan gill parasites. *Chilodonella sps, Trichodina sps; Ichthyobodo necator* 2. Gill trauma : Chemical damage. 3. Viral diseases : EHN, Hepesvirus salmonis. 4. Bacterial gill diseases: *Cytophaga sps, Flexibacter sps; Flavobacterium sps.* 5. Other bacterial diseases : *Pseudomonas sps, Yersinia sps ruckeri.* 6. Epitheliocystis infection. 7. Fungal gill infections: *Saprolegnia sps Achlya sps, Branchiomyces sps.* 8. Copepods parasites of gills: *Ergasilus sps; Salminicola sps.* 9. Inadequate nutrition (lack of vitamins).
Fish restless, congregating around pond or dam edge	1. External protozoans: *Chilodonella sps, Trichodina sps; Ichthyobodo necator, Trichophora sps.* 2. External monogenetic trematodes: *Gyro-dactylus sps; Dactylogyrus sps.* 3. Bacterial diseases: *Vibrio sps; Yersinia ruckeri, Clostridium botulinum.* 4. *Amoebae: Paramoeba sps; Thecamoeba sps.*
Flashing, i.e., twisting onto side and exposing abdomen.	1. External protozoans: *Ichthyophthirius sps, Amyloodinium sps.* 2. Internal protozoans. 3. Helminth infections. 4. Argulus infection 5. Virus diseases: IPN
Scraping bodies against bottom and sides of container	1. External protozoans: *Ichthyobodo necator, Ichthyophthirius multifilis, Amylodinium sps,* 2. External monogenetic trematodes: *Gyro-dactylus sps; Dactylogyrus sps, Cledodiscus sps.* 3. External crustaceans: *Argulus sps; Lernaea sps,*

{cont.}.....

Sign	Possible causes
Whirling along a horizontal axis	1. Virus diseases: IPN, EHN, CCV 2. Bacterial diseases: *Edwardsiella ictaluri* 3. Protozoan diseases: Hexamita. 4. Inadequate nutrition: Pyridoxine, Thiamine deficiencies 5. Aflatoxicosis 6. Other toxic chemicals : Pesticides, Herbicides
Erratic swimming movements	1. Virus diseases: CCV, EHN 2. Bacterial diseases: General 3. Argulus sps 4. Aflatoxicosis 5. Botulism 6. Toxic chemicals : Pesticides, Herbicides 7. Congenital deformaties
Swimming on side and exposing ventral side to the surface	1. Swim bladder stress syndrome 2. Gas bubble disease
Reduction of feeding activity	1. Virus diseases in general 2. Bacterial diseases in general 3. Severe parasitic infection 4. Poor water quality
Loss of weight	1. Virus diseases in general 2. Bacterial diseases in general 3. Severe parasitic infection 4. Poor water quality 5. Poor food quality
Fish on bottom of pond, fins folded, not feeding	1. External parasites 2. Internal parasites 3. Poor water quality
Fish riding high in water, dorsal surface exposed	1. Bacterial disease: *Cytophaga sps, Flexibacter columnaris* 2. Epitheliocyctis infection 3. Swim bladder stress syndrome 4. Gas bubble disease

GROSS EXTERNAL EXAMINATION

Abnormalities in external appearance of fish, should be noted and recorded using a hand lens or stereo-microscope, as well as examination by eye. The skin is then examined noting colour, swellings or visible lesions. Skin scrapes should be taken for direct wet mount examination and for fixed smears. After removing the operculum, the appearance of gills should be examined for abnormalities of colour, structural changes and large parasites. Whole gill filaments or arches should be removed for later examination under low and high power magnification. Inspection of the oral cavity should reveal any petechiae or severe periocular inflammation. Eyes which show evidence of parasites should be removed and dissected, any other abnormalities or deformalities should also be recorded. Other gross deformalities or abnormalities of fins, spine or anus may also be significant.

TABLE-10.3: EXTERNAL SIGNS OF DISEASE

External appearance	*Possible disease*
Exophthalmos	1. Many virus diseases: IPN, IHN, SVC 2. Bacterial kidney disease of salmonids 3. Mycobacteriosis 4. Nocardiosis 5. Enteric red mouth (*Yersinia ruckeri*) 6. Other bacteria: *Pseudomonas sps, Edwardsiella sps, Aeromonas hydrophila, Staphylococcus sps, Flavobacterium sps.* 7. Fungal diseases: Internal–*Exophiala sps* and other fungi. 8. Protozoan parasites: Internal *Hexamita sps.* PKD. 9. Nutritional deficiency: Vitamins A, C and E, Choline, Pyridoxine. 10. Kidney malfunction–Myxosporidians: *Hoferellus sps; Microsporidians: Pleistophora sps* 11. Gas bubble disease.
Eye haemorrhages cornea	1. Enteric red mouth (common) 2. Other bacterial diseases (less common) 3. Some viral diseases
Eye haemorrhages anterior chamber	1. Bacterial diseases: *Streptococcus sps., Vibrio sps.* 2. Other haemorrhagic bacteraemias 3. Haemorrhagic viral infections: VHS.SVC 4. Nutritional deficiency: Vitamin A, Riboflavin.
Corneal opacities	1. Nutritional deficency: thiamine, riboflavin 2. Kertitis due to infections agents 3. Keratitis due to toxins: thioacetamide phosphate esters 4. Trauma
Lens opacities	1. Nutritional deficiency: essential amino acids

{cont.}.......

	(except lysine) vitamin A, riboflavin, Zinc. 2. Maladaptation to seawater (Salmonids) 3. Parasites - digenetic trematodes: *Diplostomum sps.*
Dark body colouration over whole body	1. Virus diseases: chronic VHS 2. Mycobacteriosis. 3. Nocardiosis. 4. Other bacterial diseases: Enteric red mouth *Cytophaga sps; Streptococcus sps; Vibrio sps* 5. Nutritional deficiency –Vitamins: biotins pyridoxine folic acid, riboflavin, inositol. 6. Truma: rough handling 7. Stress due to poor water quality.
Dark body colouration, posterior region of fish (black tail)	1. Whirling disease of trout (*Myxosoma cerebralis*) 2. Nutritional deficiency; lysine 3. Lead poisoning 4. Aflataxicosis 5. Trauma to spine or trunk muscles
Dark body coloration, crainal area	1. Trauma
Pale body colour	1. Bacterial gill disease: *Cytophaga sps* 2. Mycobacteriosis 3. Nocardiosis 4. *Edwardsiella sps* 5. Nutritional deficiency; Choline, thiamine, fatty acids
Noticeable bright body colour	1. Mycobacteriosis, particularly in mature adult pacific salmon
Red body colour over various areas	1. Virus diseases: IHN 2. Bacterial diseases: *Aeromonas sps., Vibro sps. Speundomonas sps, Flexibacter columnaris Mycobacterium sps., Nacardia sps; Streptococcus sps.* 3. External protozoans: *Epistylis sps., Trichodina sps, Piscinoodinium sps.* 4. Helminth parasites–Monogenetic trematodes Gyrodactylus sps. –digenetic tremadodes: heavy infestation of metacercaria 5. Excess pigment from artificial feed 6. Nephrocalcinosis 7. Shock (circulatory collapse) 8. Spawning rash on operculum particularly cyprinids.

{cont.}......

Abdomen distended, including oedema (Dropsy)	1. Virus diseases. 2. Bacterial diseases: *Aeromonas sps Edwardsiella sps., Lactobacillus sps., Pseudomonas sps., Staphylococcus sps., Streptococcus sps.,* 3. Fungal infection: *Exophiala sps; Ochrocomis sps., Phoma sps.* 4. Kidney malfunction–myxosporidians; *Sphaerospora sps., Mitrospora sps.,* PKD-Microsporidians: Pleistophora sps. 5. *Cestodes: Bothriocephalus sps. Trianenophorus sps., Ligula sps.* 6. Nutritional imbalance: inositol, niacin, pyridoxine, thaiamine, vitamins A and E. 7. Nephrocalcinosis 8. Over feeding 9. Normal gonad development
Goitre-swelling in the isthmus region	1. Iodine deficiency 2. Tumour 3. Isopod infestation
Gills clubbed and abraded	1. External protozoa: *Chilodonella sps., Trichodina sps*, possibly gill amoebas 2. Bacterial gill disease *(Cytophage sps. Flavobacterium sps)* 3. Columnaris disease *(Flexibactor columnaris)* 4. Vibriosis 5. Branchiomycosis 6. Nutritional deficiency: pantothenic acid 7. Excess suspended solids 8. Excess ammonia (chronic)
Gill colouration. (A) Pale pink or white or mottled, or petechiae	1. Viral infection 2. Bacterial infection 3. Trematodes–monogenetgic: *Dactylogyrus sps* digenetic: *Sangicola sps* 4. Anoxia-flared opercula present 5. Anaemia nutritional deficiency, trauma 6. Chemical damage; very acidic water
(B) Normal colour with petechiae	1. Viral infection 2. Bacterial infection 3. Monogenetic trematodes: *Dactylogyrus sps.*
(C) Bright red–may or may not haemorrhage handling	1. Viral infection 2. Bacterial infection-vibriosis 3. Anoxia-flared opercula present

{cont.}.....

(D) Yellowish colour on gill surface	1. Protozoan parasites: *Piscinoodinium sps. Glenoodinium sps* 2. Iron salt precipitation, particularly in acidic waters
(E) Brown colour	1. Methaemoglobin (excess nitrites in water) 2. Fungal infection: *Branchiomyces sps.*
Gill filaments with portions missing	1. Bacterial damage: columnaris disease 2. Fungal damage: *Branchiomyces sps.* 3. Crustacean damage: *Salminicola sps.* isopod infestation. 4. Protozoan damage: Myxosporidians 5. Highly acidic or alkaline water. 6. Trauma: animal attacks including other fish
Gill filaments with whitish swellings (round to elongated)	1. Hycobacteriosis 2. Nocardiosis 3. Edwardsiellosis 4. Myxosporidian infestation: *Henhneguya sps., Myxobolus sps.* 5. Microsporidian infestation; *Dermocystidium sps.* 6. Glochidia infestation
Ulcers shallow	1. Bacterial haemorrhagic septicaemia *(Aeromonas hydrophila)* 2. Furunculosis (*Aeromonas salmonicida*) 3. Gold fish ulcer disease *(Aeromonas salmonicida sub sp nova)* 4 *Vibro sps.* 5. *Lactobacillus sps.* 6. Mycobacteriosis 7. Nocardiosis 8. Bacterial kidney disease of salmonids 9. Columnaris disease 10. Streptococcicosis 11. Fungal infections: *Saprolegnia sps. Achlya sps.* 12. Epizootic infections haematopoietic necrosis virus 13. Mycrosporidian: *Pleistophora sps* 14. Epistylis infestation 15. Nematode infection: *Philometroides sps* 16. Epistylis infestation 17. Nematode infection: *Philometroides sps.* 18. Nutritional deficiency: vitamin C 19. Epizootic ulcerative syndrome (possible-mycoses) 20. Trauma; mammal damage, bird damage, Lamprey damage, leech damage

{cont.}.......

Ulcers-deep	1. Gold fish ulcer disease 2. Furunculosis 3. Vibrio anguillarum (acute) 4. Mycobacteriosis 5. Nocardiosis 6. Bacterial kidney disease of salmonids 7. Edwardsiella tarda (chronic-deep cavities) 8. Columnaris disease 9. Epizootic ulcerative syndrome (possible mycoses) 10. Other mycotic myosites 11. Trauma-mammal damage, bird damage, lamprey damage.
White cysts in skin, fins or gills	1. Myxosporidians; *Henneguya sps., Myxosoma sps., Myxidium sps.* 2. Helminth infestations: various digenetic trematode metacercaria, nematodes 3. Fungal infection Dermocystidium.
Petechaiae on skin and in muscle	1. Developing ulclers-bacterial infection; *Aeromonas sps, Vibrio sps, Nocardia sps, Renibacterium slamoninarum, Flexibacter sps, Streptococcus sps, Lactobacillus sps, Mycobactexium sps.* 2. Protozoan parasites; *Epistylis sps.* 3. Crustacean parasites; *Argulus sps* (bite mark) *Lerneae sps (bite mark)* 4. Early fungal invasion: *Achlya sps: Saprolegnia sps*, other unidentified dermatophytes. 5. Leech damage: *Piscicola sps.*
Opaque film on external Skin surface	1. External protozoan infection: *Piscinoodinium sps. Trichodina sps, Lehthyobodo sps.* 2. External monogentic trematode infection: *Gyrodactylus sps, Benedenia sps.* 3. Toxic chemical in water, including very high or very low pH. 4. Nutritional deficiency: thiamine or biotin.
Greyish cotton wool-like fuzzy growth on skin, fins, gills, growth collapses when fish removed from water	1. Saprolegniasis.
Greyish patches on gills around mouth	1. Columnaris disease (freshwater columanaris) 2. Mycosis: *Achlya sps.,*

{cont.}......

Not fuzzy, does not collapse when fish removed from water	
White to whitish spots, round to ablong, small diameter, on skin, fins and or gills	1. Protozoan parasites: *Lehthyophthiria sps, Myeidium sps, Myxobolus sps.* 2. Virus Lymphocystis 3. Monogenetic trematodes: *Benedenia sps.* 4. Crustacean parasites: *Argulus sps.*
Black spots in skin and muscle	1. Trematodes-digenetic metacercariae: *Posthodiplostomum sps, Neodiplostomum sps, Cryptocotyle sps, Crassiphiala sps.*
Cloudiness, bluish-grey film, greyish slimy sludges on skin, fins and gills	1. Chilodoniasis 2. Trochodinasis 3. Costiasis 4. Cryptobiasis (gills only) 5 Nutritional deficiency: biotin, pyridoxine
Blue grey patches (large) on the skin of large salmonid fish	1. Nutritional deficiency: biotin, pyridoxine
White to grey mucus covering the skin, looping type motion visible through a hand lens	1. *Dactylogyrus sps.* 2. *Gyrodactylus sps* (skin, fins, rarely gills)
Body swellings	1. Tumours 2. Myxosporidians: *Ceratomyxa sps.* 3. Mycrosporidians: *Pleistophora sps.* 4. Some bacterial diseases: Furunculosis, Nocardiosis: Mycobacteriosis 5. Helminth cysts: *Eustrongyloides sps.*
Fins with clear to white swellings	1. Lchthyophthiriasis 2. Argulus sps. 3. Glochidia infestation.
Fins with spiky, naked appearance of fin rays	1. Bacterial infectiong: *Flexibacter sps., Cytophaga sps., Flavobacterium sps. Aeromonas hydrophila, Pseudomonas sps.* infection 2. External protozoan infections: *Trichodina sps., Epistylis sps.*
Fins eaten away, edges levelled off	1. Cannibalism

{cont.}......

	2. Sunburn 3. Abrasion on pond edges (particularly concrete)
Spinal deformities	1. Genetic abnormality 2. Environmental stress: congenital due to temperature and oxygen extremes or fluctuations, particularly in egg development stage. 3. Toxicity: lead, cadmium, zinc, organophosphates, organochlorides, PCB 4. Nutritional deficiency: vitamins A, C, and E, tryptophan, magnesium 5. Viral diseases: IHN, EHN, Rusurez dwarf child virus. 6. Bacterial infections: Streptococcicosis, Mycobacteriosis 7. Protozoan infestation: *Myxosoma cerebrails, Myxobolus sps.,Triangula percae, Mitraspora cypxini, Kudo sps.* 8. Fungal infection; *Ichthjyophonus hoferi* 9. Induced trauma: electric shock, physical trauma
Pale spot on skull sometimes skin eroded	1. Sunburn 2. Ulcerative dermal necrosis 3. External parasites, *Ichthyobodo necator*
Inflamed, haemorrhagic and protrusion or prolapse	1. Viral entexitis: EHN, SVC 2. Bacteraemias: Streptococcicosis, *Vibrio sps., Aeromonas Sps., Edwarasiella tarda* 3. Fungal infection: *Phoma sps.* 4. Nematode infestation: *Camallanus sps.* 5. Unsuitable food 6. Irritation from chemicals in water
Trailing faecal casts	1. Viral infection: IPN, IHN, Herpeslvirus salmonis (juvenile salmonids) 2. Bacterial infection 3. Protozoan infection: *Eimeria sps., Heramita sps.*
Projections from body,	1. Copepods: *Lernaca sps, Lernacocera sps. Caligus sps.* 2. Isopods 3. Leeches: *Piscicola sps.*

The correlation between internal clinical signs of disease and nature of the possible pathological condition are as follows:

Internal clinical signs of disease		*Possible pathological condition*
Inflammation or haemorrhagins of peritonical cavity	1.	Viral infection
	2.	Bacterial infection Aerumonas diseases, Pseudomonas diseases, *Lactobacillus* sps. *Stereptococcal* and *Staphyloccocal* infections. Mycobacteriosis Nocardiosis.
	3.	Fungal infections Exophialasis
Gastro-intestinal tract, inflammation	1.	Virus infection
	2.	Bacterial infection; Enteric red mouth disease Aeromonas diseases Pseudomonas diseases Vibriosis Streptococcal and Staphyloccocal infection
	3.	Internal protozoans *Hexamita* sps. *Eimeria* sps.
	4.	Cestodes: *Trianophorus sps.*
	5.	Nematodes: *Contracaecum sps.*
	6.	Acanthocephalans
	7.	Poisoning: Toxic chemicals
Gastro-intestinal tract contents		
(a) Clear thin mucus	1.	Infections pancreatic necrosis
	2.	Enteric red mouth disease
(b) Clear mucus plug	1.	Infections pancreatic necrosis
(c) Opaque thick mucus	1.	Infections pancreatic necrosis
	2.	Nematodes : *Philometra sps.*
	3.	Acanthocephalans
(d) Yellow thick mucus	1.	Vibrosis
	2.	Enteric red mouth disease
	3.	Aeromonas (or) Pseudomonas infection
(e) Yellow blood-stained mucus	1.	Vibriosis
	2.	Aeromonas (or) Pseudomonas infection
(f) Blood-stained mucus	1.	Furunculosis
	2.	Aeromonas (or) Pseudomonas infection
	3.	Enteric mouth disease

{Cont.}..........

g) Blood-stained part digested food	1. Furunculosis
Nodules on walls of intestine	1. Coccidiosis: *Eimeria* sps. 2. Digenetic trematods 3. Cestodes 4. Nematodes 5. Acanthacephalans
Whitish membranes enveloping digestive tract and other organs	1. Chromic Bacterial infections Mycobacteriosis, Nocardiosis Bacterial kidney disease 2. Streptococcal infection 3. Lactobacillus sps. infection
Cysts, capsule or granulosa in viscera or muscle	
(a) Whitish	1. Protozoan parasites Glugea sps. Pleistophora sps. 2. Digenetic trematodes 3. Nematode infection 4. Visceral granuloma 5. Tumours
(b) Reddish	1. Nematods *Eustrongylides sps.* *Philometroides sps.*
(c) Yellowish	1. Digenetic trematode *Clinostomum sps.*
(d) Black (in muscle)	1. Digenetic trematode *Neascus sps.*
Worms in peritoneal cavity	
(a) Large whitish	1. Larval custode *Ligula sps.*
(b) White bound worms, often in viscera	1. Nematodes
c) Red bound worms some times in viscera	1. Nematodes *Eustrongylides sps.* *Philometroides sps.*

{cont.}.......

Kidney swellings		
(a) Greyish (or) yellowish hard or soft	1.	Bacterial kidney disease of salmonids
	2.	Mycobacteriosis
	3.	Nocardiosis
	4.	Furunculosis
	5.	*Pasteurella sps.*
	6.	Proliferative kidney disease
	7.	Tumours
(b) Nodules	1.	Mycobacteriosis
	2.	Nocardiosis
	3.	*Pasteuhella sps.*
	4.	Ichthyophonus infection
	5.	Proliferative kidney disease
	6.	*Ceratomyxa sps.*
	7.	Nephrocalcinosis
	8.	Tumours
Liver		
(a) Red spotting	1.	Bacterial diseases
	2.	Virus diseases
(b) White spots	1.	Mycobacteriosis
	2.	Nocardiosis.
	3.	*Pasteurella sps.*
	4.	Ichthyophonus sps.
	5.	*Ceratomyxa sps.*
	6.	Nematode infection Anisakis larvae
(c) Black spots	1.	Fungal infection Exophiala sps *Ochroconis sps.*
(d) Yellowish spots nodules or lumps	1.	Myxosporidians
	2.	Microsporidians
	3.	Digenetic trematodes *Clinostromum sps.*
	4.	Hepatomas
(e) Patchy or generalized yellowing	1.	Lipoideal degenaration
	2.	Protozoan infection *Hexamita sps.*
	3.	Myxosporidian infection
(f) Rare whitish colour	1.	Viral infection
	2.	Bacterial infection

{cont.}.....

(g) Very dark red colour	1.	Bacterial infection
(h) Greenish colour	1.	Bacterial infection *Aeromonas sps* *Pseudomonas sps.*
	2.	Blockage of bile duct
Spleen (a) Slightly enlarged	1.	Enteric red mouth disease
	2.	Viral infection: EHN
(b) Greatly enlarged	1.	Mycobacteriosis
	2.	Nocardiosis
(c) Bright red and enlarged	1.	Bacterial infection Furunculosis
	2.	Viral infection:EHN
(d) Dark red and enlarged	1.	Furunculosis
(e) White spotting	1.	Mycobacterriosis
	2.	Nocardiosis.
	3.	*Pastensella sps*
	4.	Ichthyophonus infection
	5.	*Ceratomyxa sps.*
Heart covered by white tissue	1.	Mycobacteriosis
	2.	Nocardiosis
	3.	Streptococcal infection
Inflamed pericardial cavities	1.	Mycobacteriosis
	2.	Nocardiosis
	3.	Furunculosis
	4.	Streptococcal infection
Swim bladder (a) Ruptured	1.	Explosion is water
(b) Inflamed	1.	Virus infection:SVC
	2.	Bacterial infection Mycobacteriosis Nocardiosis Streptococcicosis
	3.	Fungal infection Exophiala sps. *Phoma sps.*
	4.	Helminth infestation Nematodes

CLINICAL EXAMINATION OF DISEASED PRAWNS

The significant technical advances made over the past few decades in relation to the culture and propagation of selected species of aquatic animals have come to form the basis of a successful aquaculture industry. In the wake of these technological achievements has also come a wider appreciation of the potential significance of disease problems to the commercial viabilities of the cultured species, and of the need to recognize, present and control such problems in practice. At the same time, the need to evaluate the impact of disease problems, together with the feasibility of their timely prevention and control, forms an essential part of research and development programmes centred on one or more species for which an aquaculture potential has been duly identified. The early recognition and diagnosis of disease problems, therefore, constitute fundamental priorities in any type of aquaculture activity, since it is necessary to ascertain if the death of one or of a few individuals is an isolated event or whether it represents the commencement of an epizootic disease capable of affecting the entire population. From this it becomes apparent that the possible aetiological aspects of any deaths must be thoroughly invesigated and resolved, so that attention can be immediately focused on the implementation of suitable prophylactic or control measures designed to protect the entire stock. There should also be consideration of the overall husbandry and management practices in case modification are required in order to minimize the serious risks which the occurrence of an epizootic disease would represent to the corresponding aquaculture operation. The carrying out of careful routine checks for the presence of diseases and of their aetiological agents is an important and essential component of any programme for the monitoring of the health status of cultured populations of aquatic animals. Many of the potentially serious, and frequently very costly disease problems can be avoided in captive stocks by the correct implementation of suitable methods for the early recognition and diagnosis of such problems. It is equally important to bear in mind that a routine disease detection programme can enable several of these problems to be contained at source, and thus prevent an accidental transmission of the disease agents to other hatcheries or aquaculture facilities.

Much attention is currently being given to finfish and shell fish health inspection and certification programmes, the purpose of which is to avoid the risks involved in the transmission of the more important diseases and their aetiological agents from one site to another, or from one country to another. It is clear, therefore, that those establishments which are unable to fulfil certain minimum health requirements, or which fail to implement reliable disease detection and diagnostic procedures on site, could very easily find themselves at a commercial disadvantage, both as producers and as suppliers, in the event that their products become unacceptable in trade because of latent or undetected, disease problems at source. The owners or managers of aquaculture establishments are strongly advised to ensure that their technical staff are trained and competent to act immediately in order to detect, diagnose, and control possible outbreaks of disease among the cultured stock. Before prevention and control measures can be considered, however, it is of the utmost importance that the disease condition be correctly diagnosed. The first step in the diagnostic procedure is to carry out a comprehensive and objective examination of diseased animals or, in the case of routine inspection programmes, of representative animals from the cultured population under investigation. Such an examination must be carried out by an aquatic pathologist with adequate training and experience in the recognition and interpretation of any clinical signs which may be present, and in the correct evaluation and reporting of such signs on a clinical history from relating to each and every case being studied. A trained aquatic pathologist, familiar with the recognition and interpretation of clinical signs and abnormalities, is always an essential member of any team responsible for detecting and diagnosing disease problems in culturesd populations of aquatic animals. Fundamentally, the clinical examination of diseased aquatic animals includes a number of important and inter-related steps involving the thorough visual observation of living or moribund specimens *in situ*, and then the sacrifice and autopsy of selected individual in order to investigate the external and internal signs which may indicate the presence of disease problems. When the clinical history forms are being filled in, it becomes of paramount importance that due consideration be given to the type as environment in which the animals are cultured, aspects of the construction, design and functioning of the facilities, feeding schedules, the existence of previous disease problems and the measures taken to control these, followed by the careful recording of the various observations made during the course of the clinical inspection and investigation procedures. It must obviously be assumed that the diagnostician will have basic knowledge and familiarity with the species under investigation.

BASIC DISSECTION EQUIPMENT

The examination of shell fish and fin fish for detection and investigation of possible disease signs and abnormalities includes not only a careful visual examination of the animals while they are alive, but also a close inspection of them following sacrifice for autopsy. In adddition to the culture media and other microbiological, materials, fixatives, stains, and solutions for histopathology and parasitology, the diagnostician must always ensure that suitable number and range of dissecting instruments and similar are at hand for when the autopsy procedure is to be undertaken. The following list may be used as a general guide and can be modified in the light of the type of animal under study, which includes larvae and post-larvae of penaeid prawns, fish alevins, Juvemile fish, large broodstock, etc.

2 pairs of scissors with sharp points, one large and the other small
2 pairs of scissors with blunt points, one large and the other small
2 pairs of forceps with sharp points, one large and the other small
2 pairs of forceps with blunt points, one large and the other small
1 Spatula
3 Scalpel handles, plus suitable number of disposable scalpel blades for use with same
1 pair of artery forceps
2 Mounted needles
1 Knife
Bacteriological loops
Microscope slides and coverslips
Diamond pencil, Marker, Labels
Pipettes and rubber teats
2 Small bottles with eye droppers
Glass tubes or other for holding material
Disposable plastic syringes and needles
Disposable rubber gloves
Petridishes with covers magnifying glass or hand lens
Dissecting microscope
Bunsen burner
Wooden dissecting board, dissecting pins, nails
Physiological saline, Distilled water, Fixatives, Stains
Receptacle containing a freshly prepared aqueous solution of disinfectant.

PROCEDURES FOR THE CLINICAL EXAMINATION OF PENAEID SHRIMPS

In penaeid shrimps, the most frequently observed clinical signs occur as a result of various types of stress present in the environment in which the animals are reared or maintained. Important signs include the presence of an opacity affecting the abdominal muscle tissue, respiratory problems, darkening of the gills, delayed or protracted moulting, evidence of dart-coloured and erosive areas in the exoskeleton, fouling and overgrowth of exoskeleton with filamentous bacteria and/or protozoan epibionts, abnormal, erratic and disorientated swimming movements, changes in the clotting time of the hemolymph. The following procedures are convenient for the clinical examination of eggs, larvae, and post-larvae, juneviles and adult prawns respectively.

EGGS

Under normal circumstances, the egg masses affixed to the ventral surface of the abdomen of berried females have a colour which varies from dark-green to almost black when the eggs are extruded, and a much lighter greenish colour when the eggs have practically completed their process of embryonic development. The occurrence of eggs which show a greyish-white or orange colour, of eggs which emit a foul and unpleasant smell, or which show gross signs of obvious necrotic charges, are always assumed to be clinical signs which indicate the strong possibility of pathological problems being present. The detection of such signs, therefore, constitutes a sufficient basis to warrant a more exhaustive examination of the material.

A representative group of eggs is examined under dissecting micropose in order to detect the presence of epibionts on the egg surface. For example, filamentous bacteria of the Leucothrix mucor type, fungal hyphae, colonial ciliates such as Epistylis and Zoothamnium sps. On completion of the external examination, the outer egg membrane is cut, and with the greatest care is removed from the egg surface in a manner such as to expose the underlying primary egg membrane. This latter is similarly examined under a dissecting microscope to detect the presence of epibionts. In turn, the primary egg membrane is itself removed, so that the embryo is left exposed. Wet mounts are made of the secondary and of the egg membranes, and of the carapace and hepatopancreas of the embryo. These wet mounts are examined with a compound microscope. On examining the primary egg membrane, it is particularly important to determine whether, any surface fauna or flora detected are really epibionts or whether, indeed, they may have penetrated the membrane and reached the underlying tissues. Particular attention is given to the detection of any possible necrotic foci in the hepotopancreatic tissue. When the carapace is being examined, an observation must be made for the possible detection of fungal hyphae, *Lagenidium* sps. Which are usually found in the integument of the sub-carapace and which, if present, indicate the possibility of an early mycotic infection.

LARVAE AND POST-LARVAE

Representative living specimens of the various larval and post-larval stages of growth and development are visually inspected to detect the presence of any obvious clinical signs of disease. For this purpose, it is convenient that the specimens be maintained in aquaria. Apparently healthy individuals normally show a clear, translucent colour. The presence of grenish-white or opaque-coloured areas is highly suggestive of possible pathological changes, the nature of which require further investigation and elucidation.

The first step is to ensure that the exoskeleton is carefully examined. Should the exoskeleton possess a soft or spongy texture when touched, this may indicate a recent moult, a nutritional deficiency or an incipient microbial invasion problem. The presence of small bubbles in the gill cavity is frequently a sign of air embolism. The behaviour of the animals is observed with care, and due note is taken of any abnormal swimming movement, dorsal flexure or lethargy. Individual specimens showing apparent disease signs are selected and removed from the holding unit and placed in a petri dish containing sea water or freshwater. The animals are then examined under a dissecting microscope (× 25). The initial examination should take into account the possible presence of any small lesions or sounds. Attention is also given to the presence of protruding gill arches on the ventral margin of the carapace, a condition which could be caused by problems experienced during the moulting process.

The presence of filamentous bacteria and other epibionts can usually be detected most frequently on the dorsal surface of the carapace and on the surface of the appendages. When circumstances so require in practice, larvae and post-larvae can usually be partially tranquilized by hypothermia, for which the addition of crushed ice to the water is recommended.

When the external examination has been completed, the animals are carefully dissected. In the first place, using a pair of scissors with fine points and a pair of fine-tipped forceps, the carapace is separated from the connective tissue, and the hepatopancreas exposed *in situ*. The hepatopancreas is closely examined, and it has mushy or a watery appearance when probed with a mounted needle, there may be necrosis of the hepatopancreas tissue. Wet mounts of the gills, hepatopancreas and integument of the sub-carapace are prepared and examined with a compound microscope. It is quite common to find that larvae infected by Lagenidium sps. show masses of fungal hyphae in the integument of the sub-carapace and characteristic zoospore vesicles may also be found in this area. The gills and hepatopancreas are examined for the presence of bacterial, mycotic and protozoan epibionts and for evidence of any degenerative changes.

JUVENILES AND ADULTS

In general terms, the clinical examination of juvenile and adult penaeid shrimps follows the procedures outlined above. The larger size of the animals, however, permits a more detailed and methodical study of the various organs, tissues and appendages. Some of the more common clinical signs and abnormalities, together with their possible causes are described in the following sections.

ABNORMAL SWIMMING MOVEMENTS

Erratic or disorientated swimming behaviour is frequently one of the first clinical manifestations of certain microbial diseases (e.g., IHHN, Bacterial septicaemia) to be seen in living animals. The shrimps suddenly lose slowly to the water surface and then turn over with the ventral surface upper most; movements of the pereiopods and pleopods cease and the animals sink to the bottom of the tank or holding unit, where they lie in a lethargic manner. The rapid and often frenetic, disorderly swimming movements alternate with periods of general lethargy. This process may be repeated on several occasions, until the death of the shrimps finally occurs. In the case of IHHN, death usually occurs within four to 12 hours of the first appearance of the disorientated swimming behaviour.

SOFT OR SPONGY EXOSKELETON

The presence of a soft or spongy exoskeleton may indicate a recent moult (check for casts), a possible nutritional deficiency, or a possible latent microbial infection.

OPACITY OF THE ABNORMAL MUSCLE TISSUE

The occurrence of an abnormal opacity, usually characterized by a greyish-white colouration affecting the abnormal muscle tissue is a sign which may be associated with a number of microbial or protozoan infection of penaeid prawns. This type of sign has been reported for viral infections (HPV, IHHN infections), bacterial infections (Bacterial septicaemia due to Vibrio sps.), mycotic infections (Lagenidium sps,) and protozoan infections and infestations (Zoothamnium sps.; certain of the microsporidioses).

DORSAL FLEXURE

At times, the shrimps may show a slight to moderate dorsal flexure of the third abdominal segment. This sign has been reported from penaeids affected by bacterial septicaemia or infested by the colonial ciliate *Zoothamnium* sps.

ABNORMAL COLOURATION

In Juvenile and adult penaeids, there may be an obvious expansion of chromatophores, particularly the erythrophores, in cases of bacterial septicaemia (*Vibrio* sps.). Brownish and eroded areas on the exoskeleton, frequently commencing as small, circular spots, brown spots, possibly with white margins are indicative of "shell disease" of "Brown spot disease", an infection of mixed aetiology caused by ehitino elastic or chitinolytic bacteria and is which funge commonly intervene. The condition known as "white shell disease" is believed to be a final infection, possibly caused by Atkinsiella dubia or similar types of phycomycetes. An extensive dark blue-blackish pigmentation on the dorsal and dorso-lateral part of the shrimps frequently indicates the presence of an infection caused by *Pleistophora* sps.

FUZZY APPEARANCE OF THE BODY SURFACE

This condition is usually associated with the presence of fouling organisms and epibionts such as *Leucothrix mucor, Epislylis,* and *Zoothamnium* sps.

HAEMOLYMPH

The haemolymph, when withdrawn from the heart or haemocoel with a syringe, may appear turbid and may clot very slowly, in cases of bacterial septicaemia. In such cases, haemolymph smears stained with Giemsa's stain show obvious signs of haemocytopenia. Wet mounts of moribund animals, examined under a × 450 magnification, show large numbers of motile bacteria; when Gram-strained smears of same material are examined, these organisms are Gram-negative rods.

GILL ABNORMALITIES

Penaeid shrimps which are heavily infested by filamentous bacteria may have gills showing an abnormal yellow-brown or greenish colour, depending on the type and colour of the detritus and/or algae trapped within the trichomes of the filamentous bacteria attached to the external gill surface. The presence of large numbers of motile Gram-negative rods in the haemolymph channels of the gills is of importance in the diagnosis of cases of bacterial septicaemia in penaeid shrimps. Melanization and necrosis of the gills is frequently due to infections caused by Furasium solani or similar types of fungi Inperfecti and the detection of typical boat-shaped macroconidia in wet mounts as the gill is of definitive diagnostic importance.

FUNGI IN THE BODY

Cases of systemic mycotic infections of larvae and post-larvae invariably show the presence of an extensive and highly-branched mycelium throughout the entire body; the individual hyphae tend to have a pale yellowish-green colour and numerous refractile droplets may be observed in their cytoplasm.

VIRUSES

Finally in view of the fact that most of the virus diseases of penaeid shrimps are difficult to diagnose precisely, in the absence of suitable cell lines on which to grow the viruses, the principal clinical signs which have been reported in the literature for the more common or important viruses are briefly summarized as follows.

(a) *Baculovirus Penaei (BP)*

Quite serious mortalities occur in larvae and post-larvae; heavily infested shrimps may be lethargic, but other well-defined clinical signs have been reported.

(b) *Monodon baculovirus (MBV)*

Post-larvae of *Penaeus monodon* tend to be lethargic, do not feed well, and show a reduced preeming activity (which makes them more susceptible to surface fouling by epibionts and filamentous bacteria); light to moderate infections tend to produce post-larvae with a smaller size, and an abnormal bluish or greyish-black colour; heavily infested post-larvae are highly suceptible to secondary infection, i.e., bacterial septicaemia.

(c) *Baculoviral mid-gut gland necrosis (BMN)*

In heavily infected *Penacus japonicus* larvae, the hepatopancreas usually shows an opaque or cloudy appearance; this abnormalities is due to necrosis of the hepatopancreatic tissue.

(d) *Hepatopancreatic parvo-like virus (HPV)*

The penaeid shrimps *Penaeus merguiensis* and *P. semisulcatus,* infected specimens, do not feed, show a poor growth rate, etc.,

(e) *Infections hypodermal and haematopoietic necrosis (IHHN)*

The first clinical sign to make its appearance is usually abnormal swimming behaviour, chracterized by a rising to the water surface where the animals remain motionless, alternating with periods of lethargy at the bottom. In acute cases of IHHN the cuticle shows a rare brown or whitish colour, and the abdominal muscle tends to be opaque, the cuticle is soft and multiple foci of meranization may be observed in the hypodermis of the body appendages cuticle, and gills, the infected shrimps do not preen themselves and become susceptible to surface fouling by filamentous bacteria and protozoan epibionts.

FISH DISEASES

Like other animals, fish can be subject to many diseases particularly when farming is intensive type. The route cause for the diseases in the case of fish is the bad quality of water and its management. When the water is either polluted or condition is deteriorated, the death of all sizes of fish is rapid. But large scale mortality in the case of fish is due to parasitic infections. Health of the fish depends on a good deal on the

TABLE 10.3: PROCEDURES FOR THE EXAMINATION OF PENAEID PRAWNS

Stage in the procedure	*Steps to be taken*	*Important actions and observations*
Receipt of the 'suspect' or diseased shrimps	Prepare a detailed clinical history form to evaluate the population under investigation	Note the species, source and origin of the stock Note the stages or the development affected Note the age, length, weight and sex of the individual specimens Note any previous "problems" reported and the control measures used Note any contact with other shrimp populations in the facility
Clinical examination of respective material	(A) External examination Detect abnormal/disoriented movements lethargy, anoxia, abdominal flexure, obvious respiratory distress, possible moulting problems; detect the presence of faecel casts, their colour/consistency	
	Detect obvious lesions and/or dislocation affecting the antennae, abdomen, appendages cephalothorax, rostrum, note any signs of shell "disease"	Prepare wet mounts and Giemsa and Gram-stained smears for microscopy; obtain material for microbiology and parasitology
	Detect obvious fouling of the exoskeleton by filamentous bacteria and/or protozoan epibionts	Prepare wet mounts and stained smears for microscopy
	Detect softening of spongy texture of the exoskeleton	Prepare wet mounts and stained smears for microscopy
	(B) Internal Examination Detect the presence of necrosis and/or	Prepare wet mounts (water and lactophenol) and Giemsia and Gram

{Cont.}..........

	melanization of the gills (Black gills)	stained smears for microscopy Remove and fix material for histopathology
	Detect the presence of opacity of the abdominal muscles (milk shrimp, cotton shrimp)	Prepare wet mounts and Giemsa-stained smears for microscopy Remove and fix material for histopathology
	Detect any chalk appearance under the cuticle	Prepare wet mounts and Giemsa-stained smears for microscopy Remove and fix material for microscopy
	Detect the presence of mycotic invasion	Prepare wet mounts for microscopy
	Detect abnormal clotting and the presence of bacteria in the haemolymph	Prepare Giemsa and Gram-stained smears for microscopy
	Check the appearance, colour and consistency of the hepatopancreas	Prepare wet mounts in distilled water or physiological saline + 0.1% malachite green for microscopy (check by light or phase contrast microscopy to detect the presence of polyhedral inclusion bodies). Remove and fix the material for histopathology and electron microscopy
	Check the intestinal and rectal contents	Prepare wet mounts for microscopy

sanitary conditions prevalent in the water body concerned. The diseases of the fish may be classified as follows:

1. Viral diseases
2. Bacterial diseases
3. Fungal diseases
4. Protozoan diseases
5. Diseases caused by worms
6. Crustacean diseases
7. Environmental diseases
8. Nutritional diseases

VIRAL DISEASES

(A) Viral haemorrhagic septicaemia (VHS)

This viral infection was discovered for the first time by Scaperclaus (1941) and the same was confirmed by Zwillenberg (1965). This particular virus has got RNA genome and measures about 50 nm in diameter. This disease becomes chronic causing very high mortality rate. Initially there is considerable reduction in the growth rate of fish. The viral infection causes significant changes in the metabolic and biochemical profiles of the tissues. Among the affected organs the damage is maximum in the case of liver and kidney. The important clinical symptoms include abnormal movement and position of fish in water, swollen eyes (exophthalmos), general anaemic condition, reduced haemoglobin percentage, protruded anus, muscular edema, swollen belly, haemorrhages in the air-bladder and in the muscles, presence of neutral or alkaline liquid in the stomach, red intestine, decaying and disintegrating liver, posteriorly swollen kidneys sores on the skin, pale gills, etc.

Control measures

There is no effective treatment of VHS disease. Fish collection should be from the unpolluted areas. Infected fish should be immediately removed and buried in far away regions. High stocking densities should be avoided. Fat material in the diet should be avoided. But vitaminous food should be preferred. Select the ponds for stocking only after the disinfection process is over. All the ponds selected for stocking should possess a clean bottom.

(B) Infectious pancreatic necrosis (IPN)

It is a viral disease causing heavy mortality in the case of young trout. This virus contains RNA genome and measures 70 nm in diameter. When fish are infected they revolve on their longitudinal axis. They swim without any sense of direction and finally becomes unable to move. They become dark in colour. The eyes protrude and the stomach and the intestine is filled with a whitish fluid. The gall bladder shows signs of necrosis, both liver and spleen are pale. Acinar cells of pancreas show severe necrosis followed by rupture and release of zymogen granules.

Treatment

There is no effcetive treatment for this disease, however, Economon (1963) has suggested the use of povidoneidine which may be useful in controlling the disease.

(C) Spring Viraemia Carp (SVC)

It is caused by Rhabdovirus carpio. The fish become dark in colour. Distended abdomen, haemorrhage in gills and skin and loss of balance are important signs of this infection.

(D) Channel Catfish Virus (CCV)

It is caused by a DNA genome virus which measures 10 nm. Fish hangs vertically in water showing loss of balance. Haemorrahage of visceral organs may be seen in this disease.

BACTERIAL DISEASES

(A) Abdominal Dropsy

The most feared disease of carp is the infectious abdominal dropsy. The bacterium *Aeromonous punctata* is the primary cause of this disease. During the spring season, the belly of the fish swells by the accumulation of yellow or pink coloured liquid in the body cavity. This is the intestinal infection, however, liver and kidney are also affected. The ulcerative form is recognised by profuse ulceration of skin by the presence of blood areas on the body and the muscle is seriously affected. When the fish are infected they feel pain and unrest which can be seen by their frequent jumping. In

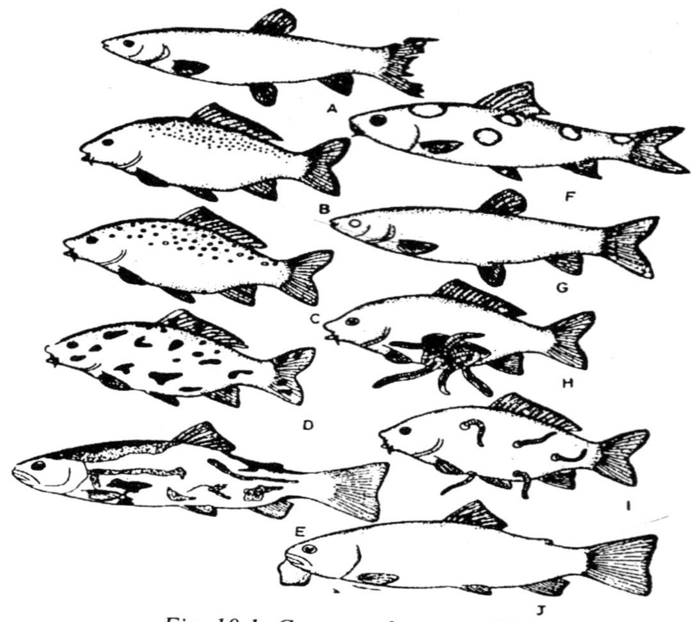

Fig. 10.1: Common diseases of fish
A. Tail rot; B. White -spot disease; C. Knot disease; D. Pox disease; E. Costiasis; F. Boil disease; G. Worm (trematodes) cataract; H. Cestode worms; I. Leech attack; J. Thyroid tumour.

some cases fins may be partially destroyed. It has been suggested that viruses are also responsible for causing this disease.

Control measures

Dropsy resistant carps should be selected for culture. Dead fish should be immediately removed. Antibiotics should be mixed either in the feed or they should be injected in a bath, to overcome any possibility of infection. Chloramphenical, Oxytetracycline and Streptophenicol are effective against *Aeromonas punctata*. For mixing the antibiotics with food, a fish with 100 g body weight should be provided 1 mg of it per day. Diping the fish in Copper sulphate solution (1:2000) and also Potassium permanganate (5 ppm) for two minutes also advisable.

(B) Furunculosis of Salmonids

This is a bacterial disease caused by *Aeromonas salmonicida*. Salmons and trouts are very easily attacked. This bacterial infection is transmitted through the digestive tract and the skin. The bacteria breed in the liver, spleen, kidneys and the blood. *Aeromonas salmonicida* is rod-shaped bacterium 2-3 um in length. Under the skin bloody boils of different sizes can be easily traced. The centre of infection is present along the muscular fibres. Inflammation of the intestines, spotted liver, haemorrhages at the bases of the fins, in the gill covers and in the muscles are also important signs of this disease. Mass mortality has been reported in trouts.

Control measures

The disease can be treated with sulphonamides, nitrofurans and antibiotics by mixing them in the food. A daily dose of 10 g Sulphamerazine and 3 g of Sulphaguanide for a period of 8 days in 100 kg of trout, 10 g of Chloramphenicol for 100 g of fry over a period of 10 days and 5 g of it for 100 kg fingerlings or trout over a period of 6 to 10 days. Oxytetracycline may also be administered. Prior to the stocking all the ponds should be disinfected with quicklime or calcium cyanide. Perfect water management in the culture ponds alleviates the bacterial infection.

FUNGAL INFECTION

(A) Saprolegniasis

Saprolegnia develops not only in the diseases or dead fish but also develop in the weakened and injured fish also. This parasitic fungus makes a way through the external lesions. Woolly grey-white or light brown bloches on the fins, eyes, gills or mouth show signs of infection. Tufts of "cotton wool" on the skin (Dermatomycosis) is a common fungal disease in several fish.

Control measures

Potassium permanganate bath 1 g per 100 lit of water for 60 to 90 minutes has been recommended or salt bath 10 g/lit water for 10 minutes or copper sulphate bath 5 g in 10 litere of water are also useful. Fungal development on trout eggs may be stopped by daily formaline bath for 15 minutes 1 to 2 ml of 30% formalin in 1 lit of water. 10 mg concentration of malachite green per litere of water for 15 minutes bath every two days also prove to be highly useful.

(B) Gill rot or Branchiomycosis

Gill rot is caused by the fungal genus *Branchyomyces* in most of the cyprinids and pike. This disease is very common in high stoking densities when there is abundance of phytoplankton and organic matter. The disease manifests during the summer season. The gill becomes pale, and on the gills gray regions can be seen. Also the gills show red coloured patches.

Control measures

Stocking density should be reduced and also supply of inorganic matter should be avoided. During hot weather even the artificial feeding should be reduced. There is a need for a good supply of freshwater. Application of 200 kg quicklime should be spread per hectare but the pH of the water is kept always less than 9. 12 kg of Copper sulpahte per hectare for pond with an average depth of 1 m has been recommended. A bath of Benzalkonium chloride 1 to 4 ppm may also be used for 1 hour.

PROTOZOAN DISEASES

Numerous protozoans can live on the external or internal parts of the fish bodies and known to cause various diseases.

(A) Costiasis

This disease is caused by a flagellate known as *Costia necatrix* when there is prolonged crow dung on tanks. They live on the skin, fins and also on the gills. Costiasis can be easily seen in salmonid fry and young carps. Slimy secretion appears on the skin, fins and the gills. Seriously affected parts of the skin show red patches. Generally affected gills are brown and partially destroyed. This disease appears when living conditions are favourable such as scarcity of food and the pH of water is acidic.

Control measures

40 ml of Formalin in 100 literes bath for 15 minutes has been advised or 10 g of Sodium chloride in liter of water for 20 minutes also prooved to be useful.

(B) Whirling disease

This disease is caused by *Myxosoma cerebralis.* Their spores present on the bottom of the pond are very resistant. The fish are infected by these spores when they are very young. In the intestine of the young fish, the spore liberates sporozoites which, when carried by the blood, settle in the soft cartilages of the head region. The growth of the fish is much poor. *Myxosoma cerebralis* produces necrosis of the bone and cartilage. Deformation of canals in the ear takes place. This causes loss of balance. The young fish whirl round and fall on the bottom. Some fishes develop black tail after infection in the sympathetic nerve. Deformation of spine, shortening of jaws and gill over, deformation of spinal column and appearance of small cavities in the head are some of the important signs of this disease.

Control measures

Storage tanks should be disinfected with quicklime or calcium chloride or benzalkonium chloride or formalin or sodium hypo chlorite.

(C) Ichthyophthiriasis

This is caused by a ciliated protozoan *Ichthyophthirius multifiliis.* The young parasites attach between the dermis and the epidermis of the fishes. This parasite can cause great losses among carp, trout and other young fishes. Small greyish spots ranging between 0.1 to 1 mm in diameter are visible on the skin of the fishes. The gills are also attacked by these parasites.

Control measures

As parasites are under the epidermis, it is not possible to get rid of the disease. Transfer of fishes from one pond to the another has been suggested. The diseased fish should be carefully removed. Application of 1500 kg quicklime / hectare with an interval of 2 days. Commercial formalin bath (1/4000) or common salt (30 g/ litre) bath are also advised.

(D) Myxobolosis (Boil disease)

The boil disease or ulceration is caused by a sporozoan, Myxobolus. Tiny knots are formed in the skin of the common carp, *Cyprinus carpio.*

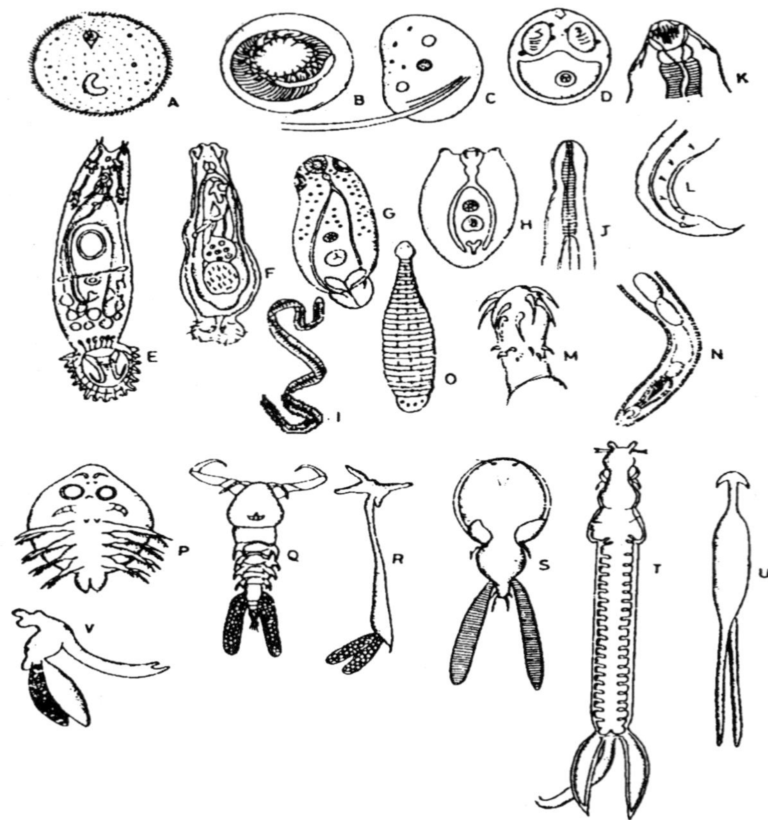

Fig. 10.2: Common parasites of fish

A. Ichthyophthirius; B. Trichodina; C. Costia; D. Myxobolus; (E-H) Trematodes; E. Gyrodactylus; F. Dactylogyrus; G. Diplostomulum; H. Diplostomum; I. Cestode, *Ligula; (J-L) Nematodes; J. Philometra (K & L) Camallanus* (anterior and posterior regions)*; (M & N)* Acanthocephalan, *Zelanechinorhynchus* (Proboscis and posterior regions); *O. Hirudinean,* Hemiclepsis; *(P-V)* Copepods; *P. Argulus; Q. Ergasilus; R. Lernaea; S. Caligus; T. Pscudocycnus; U. Lernaenicus; V. Clavellisa*

DISEASES CAUSED BY WORMS

(A) Diseases caused by Dactylogyrus

Dactylogyrus is also known as the gill fluke. It is a small flat nematode which can be great danger of young carps. *Dactylogyrus vaslator* is a common species that attacks the gills of carps. It is only 1 mm in length. A large number and variety of parasites can attack the young fish at a time. The gills swell and become grey at the edges and then are partially destroyed.

Control measures

Stocking densities should be reduced. A formalin bath 1 ml/lit of water for 15 minutes for the large fish and 0.25 to 0.50 ml/lit of water for 30 minutes for the young fish has been recommended. A salt bath of 25 g /lit of water for 10 minutes is also much effective.

(B) Piscicolosis

Fish are attacked by external parasites such as *Piscicola geometra,* the blood sucking leech.The parasite is being provided with a sucker at the end of its body, with the help of which it attaches to the body of the fish and sucks the blood. This is cylindrical, ringed worm not more than 3 cm in its length, 1 mm in diameter. The parasite makes several incisions in the body of the fish through which fungal attack is feared.

Control measures

The fish should be dipped in the Lysol 1 ml in 5 lit of water for 5 to 10 seconds. Then they are rinced in the freshwater immediately. Lysol is a mixture of cresol (50%) and soap (50%). Sodium chloride bath 10 g in 1 lit of water for 20 minutes also been recommended.

(C) Ligulosis

Ligulosis is caused by yellowish-white tape worm known as *Ligula intestinalis.* This tape worm is about 15 to 45 cm long and 0.6 to 1.5 cm wide. This disease can be seen frequently among cyprinids found in open water. This parasi . is capable of multiplication and constitute half the total weight of the host fish. The disease can be prevented by eliminating the water foul.

(D) Diplostomosis

This disease is caused by trematode, *Diplostomulum.* "Parasitic Cataract" is formed in the eye of the fish by the destruction of the lens and as a result, vision is lost.

CRUSTACEAN DISEASES

Argulosis

Argulus is a crustacean parasite of fish. Its colour is green-yellow. This parasite attaches to the skin of the fish at the base of the fins by means of hooks and suckers situated under the eyes. Anaemia may result because of wounds, subsequently leading to the death of the fish. *Argulus* discharges toxic substance from the poison gland, which is very harmful for the host fish.

Control measures

The fish should be dipped in the Lysol 1 ml in 5 lit of water for 5 to 10 seconds. Then they are rinced in the freshwater immediately. Lysol is a mixture of cresol (50%) and soap (50%). A Potassium permanganate bath 1 g in 1 lit of water for 40 seconds has also been recommended.

ENVIRONMENTAL DISEASES

The physico-chemical properties of pond water are very significant in fish culture. Number of cultured fish are very sensitive to the physical and chemical deficiencies of water.

The oxygen requirement varies in different fish. The oxygen deficiency causes suffocation in water for the fish and mortality can start. Therefore water exchange is the best possible way to overcome the oxygen debt.

During fish culture the pH of the pond water should be kept always below 9.0, otherwise it will be again toxic for the fish.

Similarly temparature variation is also harmful for the fish culture activity. Being a poikilotherm the fish can not withstand maximum variation in the temperature.

At some times the pH of the culture water goes below 5.5, the water becomes gradually toxic to fish. The whitish film covers the body of the fish and mucous is secreted. In such cases 500 kg of calcium carbonate should be spread in one hectare of the pond so that the desired pH of the water can be obtained.

NUTRITIONAL DISEASES

Overfeeding may cause several diseases, which may ultimately lead to the death of the fish. Lipoid hepatic degeneration of Salmonids. Enteritis in some fish and hepatoma or rainbow trout are all nutritional diseases. The controlling measures including the avoidance of excess food. The food should be rich in all the essential ingredients which are basically required for the optimum growth of the fish. One day fasting in every week has been recommended.

PRAWN DISEASES

Prawns are a valuable export oriented commodity. Potentiality of prawn farming to develop into an economically successful industry with high returns on investment is recognized and attracted a large number of investors in different parts of India. With the availability of different culture technologies, prawn farming is undergoing rapid intensification. While it is true that income from intensive systems is high, there is a greater risk of disease occurrence. The development of the industry has been accompanied by the study on causes of occurrence of disease of infectious and non-infectious type. Possibility of incidence of disease is high in semi-intensive and intensive culture systems. Except for certain types of parasitic diseases, it is very nature of semi-intensive culture system that encourages the development and transmission of many prawn diseases.

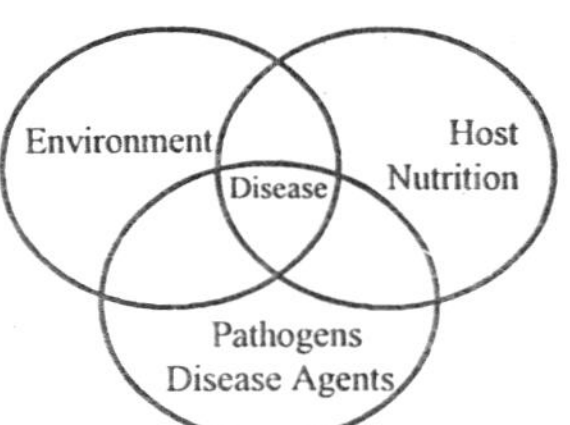

Fig. 10.3: Development of disease

Prawns, as other aquatic organisms are susceptible to diseases which cause mortalities, retarded growth and unacceptable looking prawns which alarmingly reduce profit from the prawn farming projects. Considering the significant impact, it is necessary to prawn farmers to understand the biological and environmental factors that lead to disease manifestation and maladies that can cause considerable losses among cultured stocks.

DEVELOPMENT OF DISEASES

In the development of disease on prawns as a host, three factors interact, e.g., the environment, the nutrition and the pathogens. Each of these alone or two of them acting together or all three of them together can induce stress, and consequently cause disease.

(A) The shrimp host

Generally the shrimp body is being covered by a specific exoskeleton, which is regularly replaced by a new one by the process of moulting. The moulting process increases the energy requirement of the prawns and renders the shrimp suscetible to disease agents or cannabalism.

(B) Environment

The environment of the shrimp consists of the pond soil, rearing water and the organisms in it. The growth of the cultured shrimps is influenced by several parameters including pH, dissolved oxygen, temperature, light, ammonia, nitrite, etc. All these factors affect the condition of the shrimp and the disease agents.

(C) Disease agents

The diseases are being caused either due to stress factor or infectious agents such as virus, bacteria, fungi and parasites, which may be a part of the rearing water or pond soil. If the disease causing agent finds the environment favourable for its multiplication, its population increases to overwhelm the defence of the host organism, which may result in either a diseased condition or death of the host.

(D) Immune response

Prawns show two types of immune reactions, humoral and cellular. The shrimp's hemolymph, suppose to contain dissolved substances capable of killing bacteria. These humoral substances are like antibodies, but not specific in what organism they affect. While cellular immunity is provided by phagocytes, that ingest bacteria, thereby preventing their spread within the animals body. So, both humoral and cellular systems are vital to shrimp survival and they may be only way the animals can combat disease.

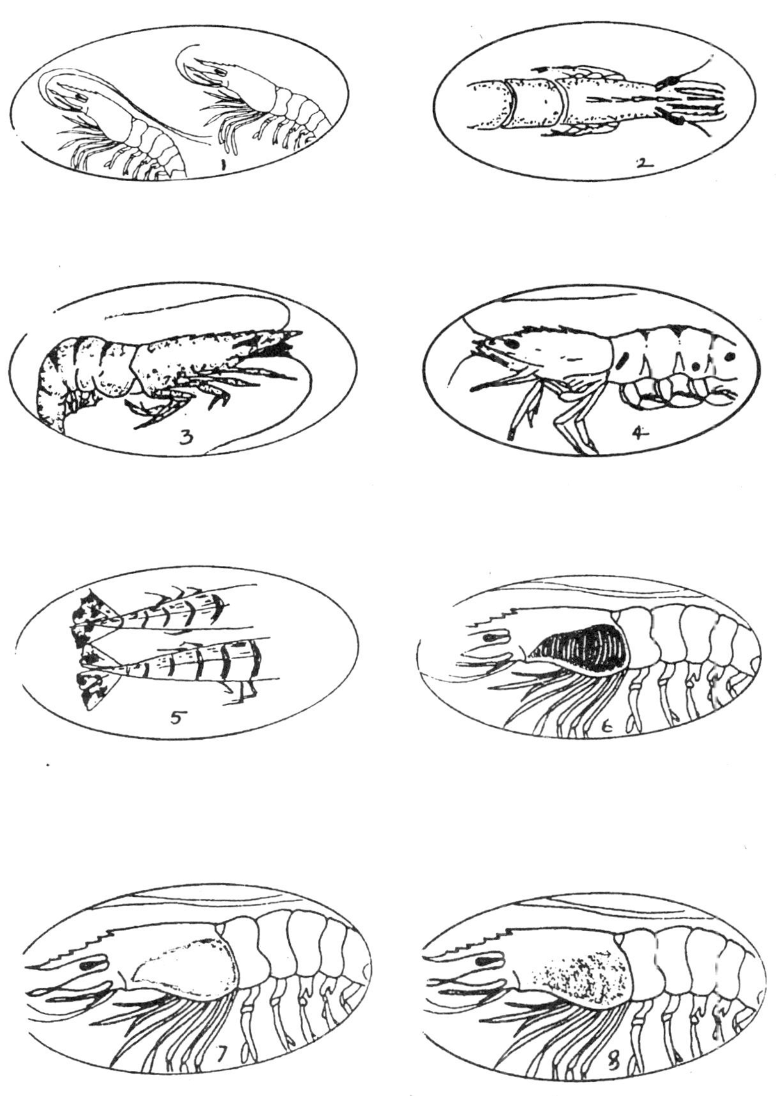

Fig. 10.4: Comon disease of prawns

1. Antennular Rot, 2. Hard Shell/Moult rotordation, 3. Red Disease, 4. Black and White Spot, 5. Tail Rot, 6. Black Gill, 7. Red Gill, 8. Brown Gill

(E) Human factors

Number of human factors are also important in shrimp farming, such as handling of post larvae transport from hatchery to the pond site for culture. Planning and implementing sound farming management strategies consisting of proper pond preparation and water management to prevent adverse environmental conditions, correct feeding and feed storage, are the basic responsibilities, which depend entirely on the person who looks after the farm management.

Detection of disease outbreaks

The following important signs help for early detection of disease in ponds.

Loss of appetite
Change in body colour
Lethargy
Abnormal swimming
Slow growth and no moulting
Antennular rot
Exoskeletal rot
Physical deformity
Opaque musculature
Empty intestinal gut / Abnormal faeces
Abnormal condition / Colour of gills
Exoskeletal epibionts
Mortalities

Onset of diseases can be detected early by careful observation of shrimps in ponds. Gradual increasing mortalities over several days to weeks usually indicate disease caused by pathogenic organisms such as virus, bacteria, fungi, ecto or endo parasites or nutritional defeciencies (Vitamin C, Calcium, etc.). Sudden mass mortalities are generally associated with adverse environmental parameters such as depletion of dissolved oxygen, accumulation of toxic substances, etc.

Diagnosis

Increase of disease outbreaks, without any characteristic manifestation, the shrimp farmer can narrow down the possible cause of the disease by the mortalitity pattern, i.e., gradual increasing mortalities over several days to weeks are usually caused by microorganisms, such as virus, bacteria, fungi, parasites or nutritional defeciencies. Sudden mass mortalities are generally associated with adverse physico chemical parameters such as low dissolved oxygen, acidic soil, high ammonia, high nitrite levels, thermal, salinity fluctuations, toxic substances.

Correct diagnosis is essential for selecting the best possible treatment for the disease and for reference for future disease combating practices.

VIRUS DISEASES

Viruses are the most difficult to diagnose. Confirmation of the diagnosis is usually made through Electron Microscopy. The viruses so far reported in cultured penaeids are

1. MBV : *Monodon baculovirus*
2. BP : *Baculovirus penaeid*
3. BMN : Baculoviral midgut gland necrosis
4. IHHNV: Infectious hypodermal and hematopoietic necrosis virus
5. REO : Reo-like virus in hepatopancreas
6. HPV : Hepatopancreatic parvo-like virus

The virus that had caused large scale mortality of shrimps in the culture farms all along the coast of Andhra Pradesh has been identified as China virus because the said virus first made its appearance in China in 1993. This is also called as Systemic Ectodermal and Mesodermal Baculovirus (SEMBV).

MONODON BACULOVIRUS (MBV) DISEASE

Causative agent : *P. monodon*-type Baculovirus

Species affected : *Penaeus monodon, P. merguiensis*

Stages affected : Post-larvae, Juveniles, Adults

Gross signs

- Affected shrimps exhibit pale bluish-grey to blue-black colouration.
- Sluggish and inactive swimming movements

- Loss of appetite
- Retarded growth
- Increased growth of benthic diatoms and filamentous bacteria may cause fouling on the exoskeleton
- Yellowish-white hepatopancreas

Effects on Host

- Presence of virus causes destruction of the hepatopancreas and lining of the digestive tract. Damage in these organs could weaken the shrimp and lead to gradual mortalities.

- Accumulated mortality of 70% was observed among *P. monodon* juveniles

- Spherical, eosinophilic occlusion bodies of the virus fill up enlarged nuclei of the hepatopancreatic cells or may be found in the lumen after have been destroyed.

Preventive methods

- Use MBV-free stocks
- Reduce stress by use of good husbandry practices and proper nutrition
- Destroy infected prawns by burning or burying in pits lined with lime
- Disinfect rearing facilities

Treatment : None reported.

INFECTIOUS HYPODERMAL AND HEMATOPOIETIC NECROSIS VIRUS (IHHNV) DISEASE

Causative agent :Unclassified virus probably belonging to Picornavirus

Species affected . *Penaeus monodon*

Stages affected : Post-larvae, Juveniles, Adults

Gross signs

- Prawns show erratic swimming behaviour, rising slowly to the water surface, hanging and rolling over until the ventral side is up.

- The motion of the pleopods and pereopods ceases and the animal sinks to the bottom.

- Prawns would eventually right themselves up, become weak and lose their appetite for food. They repeat the process of rising to the surface and sinking until they die usually within 4-12 hours.

Effects on the Host

- Larval stages (Zoea and Mysis) of *P. monodon* are pressumed to be latently infected.

- Presence of the virus can cause death of the cells of the cuticle, blood forming tissues and connective tissues of the shrimp.

- Death of the cells in these tissues can cause abnormal metabolism which eventually leads to mortalities.

LUMINOUS BACTERIAL DISEASE

Causative agent: *Vibrio harveyi, V. splendidus*

Species affected: *P. monodon, P. merguiensis, P. indicus*

Stages affected : Eggs, Larvae, Post-larvae.

Gross signs

- Larvae become weak and opaque-white

- Heavily infected larvae exhibit a continuous greenish lumine–scence when observed in total darkness. When viewed under the microscope, the internal tissues of these larvae are packed with highly motile bacteria

Effects on Host

- Systemic infectious result in mortalities in larvae and Post-larvae, reaching up to nearly 100% of affected population

Preventive methods

- Prevent the entry of luminous bacteria into the hatchery system by using ultraviolet-irradiated water or by employing a series of filtration equipment such as sand filters, filter bags, cartridge filters, etc., and chlorination procedures.

- Adhere to strict sanitation procedures prior to and during the larval stages of growth.
- Use only previously chlorinated water during spawning and rearing to ensure a clean environment for newly hatched and developing larvae.
- Siphon out sediments and debris from the tank bottom since these could serve as substrates for bacterial growth.

Treatment

- Water change must be 80 to 90% replacement daily.

SHELL DISEASE

Also called as Brown / Black spot, Black rot / Erosion, Blisters, Necrosis of Appendages.

Causative agent : Shell-degrading bacteria belonging to Vibrio, Aeromonas and Pseudomonas groups.

Species affected : *P. monodon, P. merguiensis, P. indicus.*

Stages affected : Larvae, Post-larvae, Juveniles, Adults.

Gross signs

- Appearance of brownish to black erosion of the carapace, Abdominal segments, rostrum, tail, gills and appendages.
- Blister containing cyanotic gelatinous fluid may develop on the carapace and abdominal segment. The blister may extend to the underside of the ventro-lateral section of the carapace creating a bulge on the underside.
- In larval and post-larval stages, the affected appendage shows a cigarette butt-like appearance.

Effects on the Host

- Infection is usually initiated at sites of punctures or injuries made from either telson or rostrum, cracks on the abdominal segment from sudden flexure of the shrimp body or from other damage caused by cannibalism.

- The affected shrimp becomes susceptible to cannibalism or dies from stress or energy exhaustion.

Preventive measures

- Maintain good water quality
- Keep organic load of the water at low levels by removing sediments, especially dead shrimps and moulted exoskeletons which harbour high numbers of bacteria on the lesions.
- Provide adequate diet.
- Minimise handling and avoid overcrowding.

Treatment

- Induce moulting as the condition is eliminated upon moulting except when underlying tissues are damaged.

FILAMENTOUS BACTERIAL DISEASE

Causative agent : *Leucothrix* sps.

Species affected : *P. monodon*, *P. merguiensis*, *P. indicus*.

Stages affected : Larvae, Post-larvae, Juveniles, Adults.

Gross signs

- Presence of fine, colourless, thread-like growth on the body surface and gills as seen under microscope.

Effects on the Host

- Infected eggs show a thick mat of filaments on the surface which may interfere with respiration and hatching.
- In larvae and Post-larvae, filamentous growth on appendages and the body surface may interfere with normal locomotory process and with molting, and may entrap other microorganisms like fungal spores, which may initiate a new infection.

- Larval shrimps are less prone to infection by filamentous bacteria than post-larval, juvenile and adult stages due to the rapid succession of molts throughout the larval stages. Frequent moulting does not allow adequate time for the bacteria to accumulate on the exoskeleton.

- In larger shrimps, filamentous bacteria on the gills and other body surfaces may result in respiratory distress at the point of attatchement. An indirect effect of such filamentous growth on the host entrapment of algae and debris which interfere with respiration and promote further fouling.

- Mortalities are due to direct effects of filamentous bacteria have been reported.

Preventive measures

- Maintain good water quality with optimum dissolved oxygen levels and low organic matter levels.

Treatment

- Cutrine Plus at 0.15 ppm copper in 24 hr flow through treatments or at 0.5 ppm copper in 4 to 6 hr static treatments for post-larval 2 or older.

LARVAL MYCOSIS

Causative agents : *Lagenidium callinectes*, *Lagenidium* sps *Haliphthoros* sps., *Sirolpidium* sps

Species affected : *P. monodon*

Stages affected : Eggs, Larvae, Early Post-larvae.

Gross signs

- Infected eggs, larvae and Post-larvae appear whitish, become weak, and may eventually die.
- Signs are readily apparent when the disease is already widespread.

Effects on the Host

- Heavy mortalities up to 100% within 2 days may occur. The fungal hyphae replace the internal tissues of the shrimp and extend outside the shrimp body to form discharge tubes. Infected eggs do not hatch and larvae lose equilibrium and exhibit respiratory difficulties.

Preventive measures

- Siphon sediments and dead shrimps.
- Reduce stocking density.
- Increase water circulation.
- Disinfect materials and tanks with 100 ppm detergent for 24 hr.

Treatment

- Treflan or trifluralin at 0.2 ppm for 24 hr.

PROTOZOAN DISEASES

Protozoan infection

Causative agents : Vorticella, Epistylis, Zoothamnium, Acinata and Ephelota.

Species affected : *P. monodon, P. merguiensis, P. indicus*

Stages affected : Eggs, Larvae, Post-larvae, Juveniles, Adults.

Gross signs

- Fuzzy mat on the shell and gills of heavily infected juveniles and adults.
- Reddish to brown gills.

Effects on the Host

- Microscopically protozoans may be observed attached to any external part of the shrimp.
- The protozoans cause locomotory and respiratory difficulties when present in large numbers on the appendages and gills, respectively, particularly at low DO levels.
- Less appetite.

Preventive measures

- Maintain good water quality.
- Avoid high organic load, heavy siltation, turbidity and low oxygen levels.

Treatment

- Among juveniles in nursery tanks, application of Chloroquin diphosphate at 1.1 ppm for 2 days.
- For Zoothamnium infection in adults 50 to 100 ppm formalin for 30 minutes.
- For Epistylis infection in juveniles 30 ppm formalin.
- Change water by draining pond and tank bottom daily to remove excess feeds, faecal matter and other organic wastes.

MICROSPORIDIOSIS

Causative agent : Microsporidia. These are endoparasitic protozoans.

Species affected : *P. monodon, P. merguiensis, P. indicus.*

Stages affected : Juveniles, Adults.

Gross signs

- Affected tissues or organs turn opaque white.

Effects on the Host

- Spores and other stages of the parasite replace the affected tissues.
- Infection rate is relatively low but the parasite is highly pathogenic.
- Infection may result in sterility of the spawners with white ovaries.

Preventive measures

- Disinfect culture facilities with chlorine or iodine containing compounds.
- Isolate and destroy infected shrimps by burning or boiling.

GREGARINE DISEASE

Causative agent : Gregarines, Protozoan parasites present in digestive tracts.

Species affected : *P. monodon*

Stages affected : Larvae, Post-larvae, Juveniles and Adults.

Gross signs

- Gregarines may be detected in the digestive tract microscopically.

Effects on the Host

- Interfere with particulate filtration through the gut.

Preventive measures

- Eliminate the molluscan intermediate host.

NUTRITIONAL, TOXIC AND ENVIRONMENTAL DISEASES CHRONIC SOFT SHELL SYNDROME

Causative agents

- Nutritional deficiency, Pesticide contamination and poor pond water quality and soil conditions.
- Insufficient water exchange
- High soil pH, Low water phosphate and low organic matter.
- Inadequate feeding practices.
- High stocking densities.

Species affected : *P. monodon.*

Stages affected : Juveniles, Adults.

Gross signs

- Shell is thin and persistently soft for several weeks.
- Shell surface is often dark, rough and wrinkled.
- Affected prawns are very weak.

Effects on the Host

- Affected prawns are soft shelled, grow slowly and eventually die
- Shrimps become more suceptible to wounding, cannibalism and surface fouling by Zoothamnium and other epicommensals.

Preventive measures

- Feed shrimps adequately and use only good quality feeds.
- Flush ponds thoroughly particularly when using chemical pesticides.
- Maintain pond water and soil of good quality.
- Proper water exchange.

Preventive measures

- Provide rigid water management, i.e., water exchange to the tune of 20 to 50% daily in ponds.
- Provide supplementary feeding.
- Check for the pesticide contamination.

BLACK GILL DISEASE

Causative agents

- Chemical contaminants like cadmium, copper, oil zinc, Potassium, etc.
- Ascorbic acid deficiency.
- Heavy siltation.
- High organic load.

Species affected : *P. monodon*

Stages affected : Larvae, Juveniles, Adults.

Gross signs

- The gills show reddish, brownish to black discolouration and atrophy at the tip of the filaments.
- In advanced cases most of the filaments are affected and gills become totally black.
- Physical deformities.
- Loss of appetite.
- Mortalities.

Effects on the Host

- Histological observation shows that the blackening of the gills may be due to heavy deposition of black pigment at the sites of heavy hemocyte activity.
- Extensive accumulation of blood cells in the gill filaments may result in respiratory disturbances.
- Secondary infection by bacteria, fungi and protozoans via the dying cells of the gills.

Preventive measures

- Avoid overfeeding.
- Avoid heavy metal discharges from the nearby factories.

- Proper water exchange.

Treatment

- Water exchange to the tune of 50 % daily for 3 to 5 days.
- If the disease is due to the Ascorbic acid deficiency, supplement diet with adequate amounts of ascorbic acid.

BLUE DISEASE

Causative agents

- Low levels of the carotinoid astaxanthin.
- Soil water quality problems.

Species affected : *P. monodon*

Stages affected : Juveniles, Adults.

Gross signs

- Sky blue colour instead of the normal brown-black.
- Prawns are lethargic.
- Sometimes soft shells may appear.

Effects on the Host

- Histopathological changes in the hepatopancreas.

Preventive methods

- Incorporate Vitamin A or carotinoid sources like yellow corn in the diet for around 45 days.
- Water exchange 10 to 15% daily.
- Reduce stocking density.
- Provide high quality feed.

Treatment

- Supplement diet with carotenoid sources.

RED DISEASE

Causative agents :

- Presence of aflatoxin in feeds.

- Associated with high inputs of lime.
- Rancid feeds.

Species affected : *P. monodon.*

Stages affected : Late Post-larvae, Juveniles, Adults.

Gross signs

- The first sign of the disease is sudden drop in feed consumption.
- The animals become lethargic and show general body weakness.
- Many animals confined to shallow waters at the pond periphery.
- Yellowish and eventually reddish discolouration of the body and appendages.
- Poor growth.
- Red, short streaks on gills.
- Reddish colour of faecal matter.

Effects on the Host

- Gradual mortalities up to 98 % in 3 months.
- Prawns are less resistant to stress.
- Necrosis of hepatopancreas.

Preventive measures

- Use only recent manufactered feeds.
- Prepare the pond bottom properly.
- Reduce the organic matter.
- Reduce lime input while pond preparation.

The viral diseases can be prevented through avoidance, sanitation and proper nutrition. Once the diseases set in, eradication should be immediately implemented.

Recommended solution for MBV epizooitic

- Establish MBV free broodstocks in order to produce MBV free post-larvae for the production farms.
- Do not raise shrimps in very high stocking density.
- Use disinfectants or chemicals to prevent the treat secondary bacterial sepsis and protozoal epicomensals, if necessary.

11

BIOTECHNOLOGY AND AQUACULTURE

The land area in agricultural sector and the water area in the aquaculture are the chief sources of food production for human consumption as well as other animals. Each of these areas can expand only at the cost of the other and not otherwise. There is a rapid reduction especially in the existing agricultural land area in the present day fast growing society due to construction of buildings by housing boards and laying out of roads, etc. Sometimes this also has an indirect effect on reducing water areas by filling them up for the purpose. At the other end is the problem as the ever growing population. It is often a challenge for the present day scientists to find solutions like increasing proportional problems of reducing productive areas on one side and the ever increasing population on the other.

Though traditional aquaculture based on empirical knowledge was confined to a few states in Eastern/North-Eastern India, modern aquaculture based on scientific knowledge is being adopted increasingly all over the country. It has already assumed the role of an organised industry in several areas and in just four decades, has come a long way from its origin as a backyard activity. Freshwater aquaculture has vast and varied resources from those mushrooming in the monsoons to those that sustain it even during the hot and dry summer and a potential that appears unlimited. From a production of 50-100 kg/ha/a in the hands of inexperienced farmers and 600-1000 kg/ha/a by experienced traditional aquaculturists to 25 t/ha/a under intensive and 250 t/ha/a under super intensive aquaculture is no longer a far cry. Several biotechnological options have already found and are finding increasing applications in aquaculture development and management.

GENETIC ENGINEERING AND AQUACULTURE

Genetic engineering provides the most modern tools for improvement of fish stocks for breeding and aquaculture through chromosome set manipulation by way of gynogenesis (all maternal) or andogenesis (all paternal) and polyploidy.

Any modifications in the natural process of living organisms can be termed as bioengineering and the technology that is developed to bring about such modifications is biotechnology. So genetic engineering is that part of bioengineering which deals with the modification of manipulation of the original genome of a given individual to bring in or provide certain advantages for a better performance in its environment. Genomic manipulations are brought about through rather tampering with the original genomic structure of an individual in a systematic manner but abide by certain biological laws. As genome refers to the chromosome complexity, it can be said as altering of the chromosome sets, may be addition of extra set is to the existing diploid complement (triploid/ tetraploid) or replacement with a duplicate set (gynogenesis/ androgenesis) of one and the same individual.

GENETIC ENGINEERING METHODS

The methods of genetic engineering can be broadly grouped under two heads, namely, genomic manipulations at chromosome level and gene transfer at DNA level.

Geneomic manipultations

There are two approaches for manipulating genes at chromosome level. One comes under indirect method and the other is direct one.

Indirect manipulations

These are carried out through conventional hybridization. Under this there are mainly three types.

(a) Intraspecific
(b) Interspecific and

(c) Intergeneric hybridization.

Hybridization brings about certain new structural changes or combinations in the genome of the resultant hybrid progeny by interfering indirectly at the gamete level.

(a) *Intraspecific hybridization*

In this, two different races or strains of the same species are crossed. Some times a domesticated individual is crossed with a wild variety of the same species as in the case of the common carp (*Cyprinus carpio*) in which cultivated common carp has been crossed with wild carp from the River Amur. The resultant hybrid is called as "Ropsa carp" and is known to have a greater resistance to cold temperatures than the cultivated carp.

(b) *Interspecific hybridization*

Crossing two species belonging to the same genus as in the case of the different species of the genus *Labeo*, e.g. *L.rohita L.calbasu, L. rohita* × *L. bata, L. rohita* × *L. gonius* etc.

(c) *Intergenric hybridization*

In this, two species belonging to two different genera are crossed as in the case of the hybrid crosses among the genera *Catla, Labeo and Cirrhinus, e.g., L. rohita* × *C. catla, L. rohita* × *C.mrigala, C. catla* × *C. mrigala* and their reciprocal hybrids.

These are some of the indirect ways of interfering with the characteristic original genome of a species by bringing in the genome of another variety of the same species or different species under the same genus or of different genera. The major difference between traditional types of genetic manipulation and gene transfer is that, in the former case, larger portions of the genome are transferred whereas the latter technique allows the transfer of a single gene or a particular gene in the step.

Consequently, the resultant hybrid offspring usually possess intermediate traits to their parents. For example, the intergenetic hybrids between, *L. rohita* and *C. catla* and the reciprocal hybrids possess smaller head than *Catla* and broader body than rohu. In the cross of Chinese grass carp *(Ctenopharyngodon idella*) and silver carp (*Hypophthalmichthys molitrix*), the hybrid developed weaker pharyngeal teeth than the grass carp and its gill rakers were in no way comparable to the numerous and fused ones of the silver carp. This happened for the simple reason that the hybrid offspring received a haploid (1 n) genome from grass carp and another haploid genome from the silver carp. Consequently the expression of genome of both the parents is incomplete and of course the dominate over taking the recessive in certain traits.

Direct methods

The direct means of genomic manipulations are concerned with the natural or artificial changes brought about in the chromosome complexity of an individual which are known as gynogenesis, androgenesis and polyploidy.

Gynogenesis

Gynogenesis is the process of embryonic development with solely maternal genome and without paternal genetic input. Though gynogenesis occurs in the nature, it is normally induced.

Natural gynogenesis is reported in some members of the family Poecilidae, e.g., *Poecilla formosa* and Cyprinidae, e.g., *Carassius auratus* gibelio reproduce through natural gynogenesis, a process rather similar to parthenogenesis.

The cross between certain species of cyprinidae such as the Chinese grass carp and silver carp also resulted in 45-65% gynogenetic offsprings. Among the members of the family pleuronectidae, gynogenesis (haploid) occurs when placid or flounder eggs are fertilised by milt of halibut. Diploidy could be induced by the application of cold shock to those haploid eggs.

Gynogenesis can be artificially induced by eliminating or denaturing the genetic material (DNA) of the penetrating sperm through irradiation either by exposing to UV or gamma rays. However, such genetically inactivated sperm cannot fertilize the eggs but only activate them to develop into haploid embryos which ultimately die during the development or soon after hatching, unless diploidy is restored. Restoration of diploidy is achieved by suppressing the second division, in other words, preventing the extrusion of the second polar body. This is done either through thermal (cold or heat) or hydrostatic pressure shocks. However, the shock should be given at an appropriate time after activation, just before the extrusion of the second polar body which may generally differ from species to species. So also the intensity and duration of shock may vary. The resultant gynogentic offspring may all develop into females where maternal homogamety exists. In the case of maternal heterogamety, the progeny results in 50% females and 50% males.

Gynogenesis could be successfully induced in Indian major carps *Catla catla, Labeo rohita and Cirrhinus mrigala.* The intensity of cold and heat shocks was 12° C and 39° C with a duration of 10 min and 1 min respectively 4 min after activation

Androgenesis

Androgenesis can be termed as the reverse process of gynogenesis which usually gives rise to all male progeny. Androgenesis though occurs spontaneously is normally induced. Spontaneous or natural androgenesis was reported to occur when female common carp was crossed with male grass carp. However, the percentage of incidence was very low. Androgenesis can be induced eliminating maternal genome in a similar way as is done for the elimination or denaturization of paternal genome in gynogenesis and activation such genetically denatured eggs by normal sperm followed by the usual thermal or pressure.

Polyploidy

Like gynogenesis and androgenesis, polyploidy also occurs in nature and can be induced too. In fish, polyploidy occurs (e.g., common carp, trout, etc.), possibly as a result of chromosomal translocations. Natural or spontaneous, polyploidy usually occurs when certain spcies of fish are crossed with other species of remotely related genus as in the case of grass carp × big head carp hybrids. Most of the hybrids of this cross were reported to be triploids. The method of inducing polypoidy is same as inducing diploid gynogenesis. However, unlike in gynogenesis, polypoidy is induced by subjecting the fertilized eggs (by normal sperm) to either thermal or pressure shocks. Triploidy is induced by preventing the extrusion of second polar body while tetraploidy is induced by blocking the first cleavase in the zygote. Triploidy and tetrapolidy in rohu and tetraploidy in *Catla* were successfully induced in CIFA, Bhubaneswar.

Gene Transfer

Isolation of eukaryotic genes and introduction of these genes into living animals through the microinjection technique and production of animal lines carrying the introduced genes is another unique aspect of modern genetic engineering. This has a 2-tier approach. The first one is the isolation as a particular gene and mass producing its protein products in bacteria. The second one is manipulating gametes of fertilized ova to generate an organism that transmit its altered gene to the next generation. Individuals carrying such foreign genes are called 'Transgenic' animals.

APPLICATION OF GENETIC ENGINEERING FOR IMPROVING FISH STOCKS

Intraspecific hybridization

This method aims at evolving more productive hybrids by crossing different genotypes of females and males having the best qualities during the initial progeny testing. The different phases of intraspecific hybridization are as follows:

(a) Selection of various genotypes of the same species.

(b) Inbreeding of the selected lines.

(c) Polyallele crossing of the different lines

(d) Comparative testing of the F_1 hybrid generation

(e) Propagation of the best hybrid producing parental lines for fish seed production.

ARTIFICIAL GYNOGENESIS, ANDROGENESIS, SEX REVERSAL AND POLYPLOIDY

Gynogenesis and androgenesis are the shortcut and effective methods to build up homozygous inbred lines in a faster way. Through intraspecific hybridization of these inbred lines, good heterosis effect can be expected. For producing gynogenetic/androgenetic inbred male/ female populations, methyltestosterone (MTS)/estrogens have to be provided along with the diet to the gynogenetic/androgenetic post–larvae from the day of yolk absorption for varying periods (2-3 weeks or more) depending on the species. MTS and 17B-estradiol were found to be very effective for attaining sex reversal from female to male and male to female respectively.

Scheme for the application of gynogenesis to improve production.

The main phases of the work are as follows:

(a) Selection of healthy females with good phenotypic characters.

(b) Producing the first gynogenetic female generation (G_1)

(c) Rearing the gynogenetic fry to produce sexually mature fish as soon as possible.

(d) Selection by phenotype for the best growth.

(e) Producing the second gynogenetic female generation (G_2).

(f) After dividing the G_2 population into two parts, one part will be reared as a female group and the other part has to be oversexed for producing gynogenetic males (G_2)

(g) Rearing the G_2 and G_2 to maturity.

(h) Crossing the above G_2 and G_2 to produce the third gynogenetic generation (G_3)

(i) Sex reversion in the G_3 progeny.

(j) Producing fourth gynogenetic generation by crossing $G_2 \times G_3 = G_4$.

During all the phases, the best females and males are selected according to their constitution. The inbred populations can be used after phase five for intraspecific hybridization. After top crossing, good heterosis effect can be expected.

Practical utility

Genetic engineering is accomplished indirectly through hybridization and directly through chromosomal manipulation and gene transfer. The technique of genetic engineering can be used as a powerful tool to derive several advantages in modern aquaculture, all contributing to augment the production either directly or indirectly. Intraspecific hybridization by crossing different genotypes of females and males helps in genetically upgrading a species having improved characteristics which ultimately result in more productive hybrids. Intergeneric hybridization in certain crosses may help in combining the useful characters of two different species of two genera, as a consequence of the new gene combination in the resultant hybrids.

Through intergeneric hybridization, the following results can be expected. New or increased feeding spectrum, better growth and survival rates, better tolerance for reduced levels of oxygen (Silver carp × Big head), better quality of meat (Common carp × Silver carp, Rohu × Catla, or Common carp × Indian carps), sterile triploid offsprings (Glass carp × Big head), etc.

Gynogenesis is an effective tool to build up inbred populations in a faster way which can be use for improving the species with their useful characters through intraspecific hybridization. It is also useful to produce monosex populations. Producing gynogenetic male populations with hormonal sex reversal (G_1) and top crossing the selected males with normal females results in all female progeny which may show improved chracters. Androgenesis has got the similar applied values as that of gynogenesis.

Polyploidy, as implied in the name, leads to increased level of ploidy in a given material. An increase in the ploidy level from a diploid to a triploid may lead to sterility in most of the animals. Sterile populations have a lot of applied value. Sterility is of great advantage in fish culture as energy spent on maturation or gonadal development may be directed beneficially for increased somatic growth. This may be very useful especially in species which have a shorter maturity cycle such as common carp and tilapia. Sterile grass carp is a very effective biological agent for controlling weeds in open-water systems without the fear of its establishment as it cannot reproduce. Sterility also helps to control indiscriminate breeding in tilapia and common carp or any such species which have prolific pond breeding habit and keeps their over population under check, thus maintaining a balance in the proper growth of the fish stocked.

Coming to the latest aspect of genetic engineering, transfer of particular gene into an individual is a novel approach. It provides extra ability, if integrated and expressed in the system of the recipient organism to grow faster to resist a particular disease or adverse environmental conditions depending on the type of gene received. So far, gene transfer in fish is following the same pattern as other animals. Development of transgenic fish will take a longer time. Attempts are being made to isolate fish growth hormone gene and success in isolating salmon growth hormone gene is already achieved. Similarly growth hormone genes in channel catfish and common carp are currently being isolated. If these genes are successfully transferred to their respective species their growth and yield rates should improve. Some attempts, however, have been made to produce transgenic fish by introducing bovine hormone gene into loachand gold fish. It is claimed that these animals grow four times faster than the normal ones and the growth hormone gene is transmitted to the next generation.

Genetic engineering is an exciting area and entertains great hopes with its wider scope of application. Its products will immediately improve the quality of life. Genetic engineering will produce better crops in agriculture as well as aquaculture and animal husbandry. It will provide solutions to the adverse problem of reducing productive areas and increasing population. It has already done a great job in the agriculture sector through green revolution by evolving high yielding strains or varieties of paddy, wheat or cereal crops and so on. It is also expected to do the same in aquaculture too.

HORMONAL APPROACH TO FISH SPAWNING

Culture of the most commonly cultivable species of fish whether carps, cat fish or murrels and shrimps depends or at least depended till a few years ago on their availability of the seed from the wild. Fluctuations in their availability and quality ofter cause severe problems. It is obvious that dependable sources of seed would provide a stable industry, the key to which will be through control over the reproductive habits of these species. Of the two approaches to induction of spawning in fish environmental and hormonal, sometimes the latter is also combined with the former to achieve the desired results. However, the use of hormones is the biotechnological approach to the problem of fish breeding. About a dozen chemical agents are already available which intervene the hypothalamous-Pituitary-Ovary axis at various levels.

Cultivable varieties of Indian major carps *Catla catla*, rohu *Labeo rohita* and mrigal *Cirrhinus mrigala* were first spawned in 1957-58 followed by exotic silver carp *Hypopthalamichthys molitrix* and grass carp *Ctenopharyngodon idella* in 1962 using crude pitutarty extract prepared from glands collected from fully mature carps during the summer months. The technique was adopted on a large scale throughout the country to obtain quality carp seed. Glands collected from certain marine catfish and other freshwater species have also been found useful though neither is in much demand nor in use.

Spawning in Indian and exotic carps and catfish has been achieved through administration of HCG (Synaharin) but only when it was used in combination with pituitary, it reduced the latter's requirement by about half. Crude HCG powder (Summach, 30 IU/mg) obtained from pregnant female urine, is now available in India on a commercial scale and being cheap and standard, is widely used especially for silver carp which require much higher doses of carp pituitary than indigenous carps. Many scientists and farmers, however, feel that HCG alone is not effective in spawning the Indian major carps but when used in combination with carp pituitary (30% HCG: 70% pituitary) effects considerable saving as the pituitary material. The antiestrogens have been found effective in inducing spawning in several species but not so in India when clomiphene citrate was injected either alone (10-200 mg/kg) or with a threshold dose (4-5 mg/kg) of carp pituitary extract. It is likely that they have not been administered at the appropriate physiological stage as fish or excessive dosages, whch inhibit gonadotropin secretion, were used.

17 -20 -P is well known maturation inducing steroid (MIS) even at low doses of 2-3 mg/kg but in order to achieve ovulation, the fish have to be primed first with pituitary extract or GIH & LHRH or LHRH-A. 17 -hydroxyprogesterine (17 -P) is, however, able to include both maturation and ovulation and has an advantage over 17 -20 -P in being cheap and readily available. Of the various corticosteroids, Deoxycorticosterone acetate (DOCA) has been used successfully for spawining *H. fossilis* experimentally. Its efficacy also, however, improves by a priming dose of pituitary extract or GIH. LHRH is effective in inducing gonadotropin release and ovulation in fish but its superactive analogues are far more effective. The dopamine antagonists, pimozide and domperidone have also been used successfully to potentiate the ovulatory effects of LHRH-A. Dopamine acts as Gn RIF and blocking its action by antagonists greatly helps the GIH release response to LHRH-A. Pimozide and LHRH-A at 10 mg and 290 ug/kg while Domeperidone, which is ten times more potent, at 5 mg with LHRH-A at 10 ug/kg have given 95-98% fertilised eggs. Another development is the administration of LHRH-A by implantation for ensuring a long-term delivery without handling the fish repeatedly. Such pellets with cholesterol base and cellulose as binder were prepared and implanted in *C.batrachus* for obtaining advance maturity with successful results. LHRH-A implants using cholesterol pellets have already resulted in spawning the difficult to breed milk fish *Chanos chanos* at the Oceanic Institute, Hawaii. Salmon GnRH analog give in combination with pimozide has also been used recently to induce off-season spawning in Rohu in Haryana by about April. Fish primed with HCG (10 IU/kg) were given a single intraperitoneal injection of the combination after 15 days of priming and spawning observed 22 hrs after the injection. However, with an intramuscular injection of the same components, an additional dose of pituitary extract was required to achieve spawning. Fertilization was 70-80% and hatching 70%. Brood-fish suffer a severe damage due to repeated netting and handling at fish farms resulting in gonadal atresia which affects spawning adversely. Triodothyronine (T_3) is believed to restore the ovulatory response to hypophysation by enhancing the effect of gonadotopin on steroidogenesis and vitellogenesis. Thyroxine has been tried with limited success in spawning the milk fish.

HORMONAL APPROACH TO CONTROL SPAWNING THROUGH SEX MANIPULATION

Controlling unwanted reproduction in prolific breeding fish such as *Tilapia* and common carp through hormonal manipulation of sex is a modern approach in aquaculture. Administration of 17 -methyl-testosterone (17 -MT) or 17 -estradiol (17 -E_2) at 200 ppm for 131 days starting with 1-day old hatchlings has been found to result in retarded gonadal development and altering the sex ratio significantly. The 17 MT treated group produced 46% males and 54% sterile forms while the 17 -E_2 group consisted of 15% males, 42% females and 43% sterile forms. Dietary administration of 17 -MT at 300 ppm for a period of 30 days starting with 1-day old hatchlings resulted in a population of 8% males, 6% females and 86% sterile individuals but at 400 ppm for the same period 98.25%. Sterile forms were obtained with only 1.75% males. However, a full brood of sterile common carp was obtained by dietary administration of 17 -MT at 300 ppm for 50 days or 500 ppm for 45 days. Mibolerone at 75 ppm also produced sterile common carp. Rearing the fry so obtained for a period of 365 days on a hormone-free diet showed an increase of 40.64% and 46.87% in the mean final weights of the two groups. Loss of weight due to evisceration was only 5.59 to 7.42% in the hormone-treated groups as compared to 14.95% in the control fish suggesting the availability of more edible meat per unit weight from hormone-treated fish.

CRYOPRESERVATION OF MILT

The success of artificial spawning through hormone treatment depends on the availability of fully mature males and females at the same time the insufficiency of milt from males which are mature but not fully ripe or milt resorption result in failure of breeding programme. These problems can be solved either through sperm preservation or enhancement of spermiation by gonadotropin administration. Rohu, *L. rohita* sperms were first preserved successfully in Holtfreter's and Ringers solutions about two decades ago. However, no further attempts at developing techniques for preservation of fish sperms were made till recently when carp spermatozoa were preserved for about 365 days under cryogenic conditions at 196°C using besides others egg-yolk, citrate, urea egg-yolk and Ma as extenders.

BIOLOGICAL NITROGEN FIXATION AND BIOFERTILIZATION

Certain bacteria and blue-green algae can use atmospheric nitrogen and taxi it to crop plants. It should not be surprising to note that nature pays the energy bill for conversion of about 140 million tonnes of nitrogen into fertilizer nitrogen each year. Proper husbandry of micro-organisms could therefore be useful in meeting a large part of the total nitrogen demand in agriculture, increasing use is being made of nitrogen fixers or biofertilizers such as *Rhizobium, Azospirillum*, blue-green algae and *Azolla*, etc.

In India, *in situ* measurements of biological nitrogen fixation have been made for the first time at CIFA, Bhubaneswar during the last few years employing acetylene reduction technique with a gas chromatograph. These estimates ranging from 0.39-0.73 kg N/ha-m/a in water and 0.86-1.68 kg N/ha/a in the sediment media of fish ponds under different management practices are low as compared to those from temperate freshwaters. Biological nitrogen fixation and biofertilisation have been identified as priority research areas by the Institute and intense efforts are being made to increase the nitrogen fixation levels through micro-organism husbandry. Use of Azolla, Anabaena association which annually fixes about 100-150 kg atmospheric nitrogen /ha from 40-60 tonnes of biomass is being tried as a biofertilizer. Blue green algae, viz., Anabaena, Nostoe and Calothrix are also being evaluated for their growth potentials and nitrogen-fixing capabilities in fish ponds.

ENVIRONMENTAL IMPROVEMENT

Stress causes several diseases, retards or checks growth and affects feed intake and assimilation. Oxygen deficiency, carbon dioxide and ammonia build-up and metabolite accumulation are the main sources of this stress. Biological filters employing methanogenic and sulphur bacteria are being used for processing distillery effluents in a factory near Madras to bring the BOD levels below 25 ppm. Bioreets are now installed *in situ* to remove metabolites and hasten mineralisation. Oxygen levels above 5 ppm and free carbon-dioxide, ammonia and BOD levels below 20 ppm, 0.02 ppm and 25 ppm respectively are the best recommended for fish culture. Aerators or agitators take care of oxygenation and improve pond sanitation considerably. This is, however, achieved in rural ponds through introduction of common carp or raking of pond bottom so that the accumulated toxic gases and metabolites are released and bacterial activity intensified for organic decomposition and mineralisation. Studies on microbial coenoses of fish ponds under different management practices, viz., fertilization with cow dung, inorganic fertilization and bottom-raking showed high heterotrophic bacterial counts both in water (no. × 103/ml, 2.02-5.89, 0.71-2.82 and 0.51-9.00) and sediment (no × 106/g, 0.39-2.15, 0.40-2.00 and 0.43-12.27 respectively) media of ponds where the bottoms were raked.

Selective enrichment of biotic communities is brought about by provision of additional substrates for settlement of periphyton, compost manuring for cladoceran growth and green manuring for development of oligochaetes. The culture of algae such as *Scenedesmus* and *Spirulina* and zooplankters like *Moina* and *Brachionus* and their nauplii for supplementing the natural food resource in the ponds is receiving increasing attention and playing a great role especially in increasing the survival rates of the young fish and shell fish during rearing Biofiltration and recycling of wastewaters mainly employ the nitrifying bacterial population for oxidising the ammonia and nitritite forms of nitrogen.

RECYCLING OF ORGANIC MATTER

Fish production in most aquatic ecosystems is largely sustained by recycling dead organic matter rather than by consumption of net primary production. Pond fertilisation with live stock manure is a well known practice. The greatest potential of this microbial detrital production system lies in reducing or eliminating the need for costly supplementary feeds. Microbial production in detritus affords a rich source of nutrients for fish and a mechanism for rapid recycling of nutrients released upon death of plants of breakdown of waste products.

Animal wastes are often fed into fish ponds directly but various treatment measures reduce the oxygen demands. The recycling of organic matter through a process of microbial degradation called composting is a well-known traditional exercise. Though it reduces the oxygen demand and kills the pathogenic organisms and is commonly used in China and Thailand for fertilizing the fish ponds, the practice is neither common in India nor recommended for large-scale use. However, processing of organic matter or fermentation, especially of the excreta from various animals, in the biogas plants gives a much better digested end-product. These are now being used on a large scale for obtaining the semi-digested slurry, rich in nitrogen and phosphorous. The nitrogen and phosphorous values are 3-4 times higher in the slurry than in raw cow dung. Its demand on oxygen budget of the pond is low and hence results in greater plankton production. Utilization of biogas slurry as the only input in fish ponds gave yields as high as 5,000 kg/ha/a resulting in an economy of 60-70% of operational costs on fish feeds. At high densities of 7,500 fingerlings/ ha, two crops of silver carp at levels of 25% of the stocking density could be taken from the ponds. Though not provided with any aquatic vegetation, grass carp also registered an excellent growth at 2% of the total stock. The digested slurry was utilised also as a feed along with rice bran for *Heteropneustes fossilis* and common carp. Partially digested plant matter excreted by grass carp can support almost an equal biomass of other fish species either by its direct utilization as feed or as a fertilizer in the fish pond. A distinctly higher heterotrophic bacterial activity was noted in such systems. A low energy, low-cost system was developed.using aquatic vegetation as the sole input in ponds stocked at 4,000 fingerlings/ha of which 50% was constituted by grass carp alone and another 50% shared equally by each of the other five species, viz., catla *(Catla catla)* , rohu (*Labeo rohita*), mrigal *(Cirrhinus mrigala*), silver carp *(Hypopthamichthys molitrix*), and grass carp (*Ctenopharyngodon idella*). Fish productions of the order of 4-4.5 t/ha/a were obtained. Partial replenishment of the vegetable or animal protein in fish of its supplementation with algal of fungal protein has tremendous advantages in that these microorganisms contain vitamins and supplementary growth factors. Production of single cell protein from lingocellulose and other organic wastes and modification of the present day fermentation processes are needed to develop an appropriate technology for aquaculture. Either pure or mixed cultures of yeast, fungi and bacteria are used to produce single cell proteins. Yeast has been included in several pelleted feed formulations and "Pruteen" grown on methanol found to be an efficient protein feed for carp. Single cell protein from petroleum products have also been used as feed in aquaculture. Recent studies on processing of water hyacinth, *Eichhornia crassipes* have indicated a period of 2-3 weeks to be adequate for processing hyacinth with cowdung (10%) and urea (2%) when bacterial activities get stabilized after a peak. Exclusive feeding of Indian major carps and common carp with processed hyacin gave production of 1.13-2.43 kg/m^2/d with conversion ratios of 20.18-37.21 on wet weight and 2.02-3.72 on dry weight basis.

DISEASE CONTROL AND VACCINE DEVELOPMENT

Studies on bacterial and viral diseases are still in infant stage in India. Primary monolayer cell culture from kidney, ovary and air bladder have been standardised and routinely done in several institutions of ICAR. Four cell lines, viz., BB, BF_2, EPC and FHM are being maintained and subcultivated. A new cell line has also been developed from, the hepatic tumour of silver carp and found sensitive to the Epizootic ulcerative syndrome. Serodiagnostic techniques are being standardised. Certain strains of *A. hydrophila* and Edwardsiella tarda have been used for serodiagnosis. A polyvalent bacteria has been developed from five strains of *A. hydrophila* and tried for experimental vaccination of culturable carps.

BIBLIOGRAPHY

Alikunhi, K.H. 1957. Fish culture in India, Indian Council of Agricultural Research, New Delhi. *Farm Bulletin* No. 29.

Anderson, D.P. 1974. *Fish Immunology*, T.F.H. Publications, Hong Kong.

Bardach, J.E., Ryther, J.H. and McLarney, W.O. 1994. *Aquaculture–The Farming and Husbandary and Marine Organisms*, Wiley Interscience, New York.

Bhaskara Rao, D., Harshitha, D. and Sambasiva Rao, K. R. S., editors 1999. *Advanced Biotechnology*, Discovery Publishing House, New Delhi.

Breder, C.M. Jr. and Rosen, D.E. 1966. *Modes of reproduction in fishes*, Natural History Press, Garden City, New York.

Brown E.E. 1983. *World Fish Farming: Cultivation and Economics*, Connecticut, AVI Publishing Co., Inc.

Boyd. C.E. 1982. *Water quality management for pond fish culture*, Elsevier Scientific Publishing Company, New York.

Chen, T.P. 1976. *Aquaculture Practices in Taiwan*, England: Fishing News (Books) Limited, England.

Chonder. S.L. 1992. *Hypophysation of Indian Major carps*. Satish Book Enterprises, India.

Falconer, D.S. 1981. *Introduction to Quantitative Genetics*, 2nd edn., Longman, London

Halver, J.E. 1979. *"Vitamin requirements of finfish."* In: *Finfish Nutrition and Fish Feed Technology*, Vol. I. (Ed. by Halver, J. E. and Tiews, K.), Schriften der Bundes for schungsanstalt fur Fischerei, Berlin.

Harris, C.C. 1978. *Fish Farming*, Pelham Books Ltd., London.

Hepher, B. and Pruginin, Y. 1981. *Commercial Fish Farming*, John Wiley and Sons, New York.

Heut, M. 1986. *A Text book of Fish Culture-Breading and Cultivation of Fish*, Fishing News (Books), London.

Hora, S.L. and Pillay, T.V.R. 1962. "Handbook on fish culture in the Indo-Pacific region." FAO Fisheries Biology Technical Paper No. 14.

Jhingran, V.G. 1982. *Fish and Fisheries of India*. Hindustan Publishing Corporation, New Delhi.

Julian, T.W. 1974. *The Concise Encyclopedia of Tropical fish*, Octopus Books Limited, London.

Maynard, L.A. and Loosi, J.K. 1969. *Animal Nutrition*, McGraw-Hill, New York.

Pillay, T.V.R. 1990. *Aquaculture–Principles and Practices*. Fishing News (Books) Ltd. London.

Shelbourne, J.E. 1964. "The artificial propogation of fish." In F. S. Russell (Ed.) *Advances in Marine Biology*, Vol. 2, Academic Press, London, New York.

Spotte, S.H. 1970. *Fish and Invertebrate Culture*, Wiley-Interscience Publishers. New York.

Stickney, R.R. 1979. *Principles of Warm-water Aquaculture*, John Willey and Sons, New York.

Streba, G. 1966. *Freshwater fishes of the world*, Viking Press, New York.

Talwar, P.K. and Jhingran, A.G. 1991. *Inland fishes of India and adjacent countries*, Vols. I & II. Oxford and IBH Publishing Co. Pvt. Ltd., New Delhi.

INDEX